Christian Bracco

When Albert became Einstein

1895-1901

Quand Albert devient Einstein was originally published in French in 2017. This translation is published by arrangement with CNRS Éditions. © CNRS Éditions, 2017

© For the English translation, EDP Sciences, 2024

ISBN (print): 978-2-7598-3480-8 – ISBN (ebook): 978-2-7598-3481-5

Preface

One of the outstanding challenges in the history of science is to account for the sudden burst of creativity in 1905 of the young Albert Einstein, then a virtually unknown expert on electrotechnology and technical clerk at the Swiss Federal Office for Intellectual Property in Berne. Within the period of a few months, Einstein submitted to the *Annalen der Physik*, the leading physics journal of the time, five papers that have been said to have "changed the face of physics". The first of these papers reintroduced the notion of discrete light quanta into physical theory, the second established methods to determine the size of molecules and of Avogadro's number and served as a doctoral thesis for the young Einstein, the third explained Brownian motion by a novel statistical reasoning, the fourth established the special theory of relativity and the fifth introduced the equivalence of energy and inertial mass, later epitomized in the ubiquitous formula $E = mc^2$.

By all accounts, this was a true *annus mirabilis* in the history of physics like the short period in 1665–67, when the young Isaac Newton, during the Plague years, went back to his little hometown of Woolsthorpe and invented differential calculus, explained the spectral composition of light, and conceived of a general law of gravitation.

How can we give a historical account of such extraordinary feats of creativity if not by throwing up our hands, admitting defeat and reverting to invoking a notion of divine genius? If it is possible at all, any adequate historical response to this challenge depends on a careful analysis of the pertinent context. Unfortunately, in Einstein's case precious little contemporary documentation is extant as far as direct sources, correspondence, notes, manuscripts are concerned. For a long time, it seemed like the meticulous and thorough edition of the first volume of the *Collected Papers of Albert Einstein*, published in 1989, had exhausted our possible sources relevant to account for the coming of age of the young Einstein. In particular, the *love letters* between Einstein and his fiancée and first wife, Mileva Einstein-Marić turned out to be a treasure trove of information about the intellectual development of Einstein during his undergraduate studies at the Zurich Polytechnic.

But it is in the nature of editorial projects that they focus on the proclaimed author whose writings they wish to make available to an interested reader. Any relevant context provided in an editorial project is necessarily focussed and filtered. A number of historical accounts of the young Einstein's intellectual development have been published based on the publication of that first volume of the *Collected Papers*. Those accounts also introduced new pertinent context, depending on their respective point of view.

The present study by Christian Bracco presents a whole new dimension of relevant historical context for an intellectual biography of Einstein. The result of many years of research, this book illuminates the wider context of Einstein's upbringing in a world of technical entrepreneurship, economical investments, intellectual stimulation, especially in late nineteenth century Italy. Going back to Einstein's family and their social connections and embeddedness in a broader intellectual and historical context, Bracco offers a wealth of circumstantial evidence for possible influences that shaped his mind.

While Einstein's Swiss years, his studies at the Polytechnic in Zurich, where he met Mileva and his subsequent work in Berne are comparatively well researched, not much has in fact been known about the early Italian context of his youth and adolescence. But as the present study shows, there are many interesting facets of Einstein's and his family's connections with Pavia and Milano, where his father and uncle engaged in the booming electrotechnical industry of the time. Following up on his family's social, economical and industrial links to the Italian environment, Bracco presents many possible influences that would have been absorbed by a budding curious mind. A key role is assigned here to Einstein's friendship to the somewhat older Michele Besso and to his family, both on his father's and his mother's side. Among Besso's relatives we find entrepreneurs, engineers, mathematicians, artists, authors, and university professors who would have known, inspired, and supported Einstein. Although a direct influence can rarely be documented, it is more than likely that education and intellectual stimulation was provided privately in a milieu of technical industry, factory production and social relations to an upper class of late nineteenth century Italian society. Einstein's own later involvement with technical patents and patents-based industry as well as the training of his own son Hans Albert in the gyrocompass factory of Hermann Anschütz testifies to this early influence and technic-industrial culture.

If we want to understand how Albert became Einstein, we are invited here to take a step back and look at the broader context of what the world may have looked like to Albert when he was not yet Einstein.

Professor Tilman Sauer
Institut für Mathematik,
Johannes Gutenberg-Universität Mainz, Germany

Table of Contents

Introduction. 1
  1905, a "miraculous" year for Albert Einstein? 1
  Two important witnesses: Mileva Marić and Michele Besso 3
    Letters to Mileva 3
    Correspondence with Michele Besso 5
  Michele's unsuccessful plea for a history of the genesis of Albert's ideas . 7
  Editorial plan. 10

1. The Italian political and industrial context of the 19th Century   **15**
  The Einsteins in Milan and Pavia: settlements and relationships 16
    The short adventure of Einstein, Garrone & Cie (1894-1896) 16
    Famous homes and political connections 17
  Historical context: Napoleonic and *Risorgimento* influences. 19
    The Cisalpine Republic 19
    The insurrection of 1848 21
    The Risorgimento and the birth of a modern state 22
    The creation and role of the Lombard Institute. 23
  The new industrial and economic framework 25
    Unification and railways 25
    The electrical industry. 26
    Insurance and the fate of workers. 27
  The academic world and its links to the industrial sector 28
    Some examples of the creation of polytechnics (Turin, Milan,
      Rome) 29
    The Italian Electrotechnical Association 31
    Hoepli and the development of Italy 32

2. The company of Jakob and Hermann Einstein
in the international electrotechnical context – exhibitions
and reviews. **35**
  Paris 1881, the first International Electricity Exhibition 37
    The birth of a major exhibition devoted to electricity
      and its applications. 37
    Electric lighting 39
    Electric dynamos 41
    Rail transport. 42
    Medical and educational applications. 43
    The first international congress of electricians. 43

Munich 1882 and the Einsteins' involvement 44
Vienna 1883 and electric meters 46
Turin 1884 and the development of alternating current 47
Frankfurt 1891, three-phase current and the Einsteins' participation . . . 50
Electrotechnical magazines 52

3. Michele Besso and his family's role in Italian industrialisation . 55
Michele Besso (1873-1955), Albert's long-standing friend 56
   A brilliant young man 56
   Constant links with Albert 57
   An endearing personality 58
   Albert and Michele meet 60
The paternal branch of the Besso family 63
   Giuseppe Besso (1839-1901), Michele's father, humanist. 63
   Beniamino Besso (1840-1907), railway engineer and scientific
      author 64
   Marco Besso (1843-1920), the influential Chairman of General
      Insurance. 66
   Davide Besso (1845-1906), mathematician and teacher. 68
The maternal branch of the Cantoni family 70
   Vittorio Cantoni (1857-1930), civil Polytechnic engineer 71
   Giuseppe Jung (1845-1926), professor of graphic statics
      at the Milan Polytechnic 73

4. Albert's environment in Pavia and preparations for ETH 75
Albert's attempt to be admitted to ETH in October 1895, at the age
   of sixteen without a school-leaving certificate 76
   The plan to be admitted to ETH. 76
   The letter to Galileo Ferraris 77
   Intervention with Herzog 78
The scientific memoir of 1895. 79
   The circumstances 79
   The subject: the state of the ether in a magnetic field. 80
   What are the sources for the dissertation? 81
The Einsteins' links with the University of Pavia 83
   Their partner Lorenzo Garrone and the mathematician
      Giulio Vivanti 83
   Their partner Angelo Cerri, theoretical geodesy assistant 84
   Jakob Einstein, Otto Neustätter and the medical academics
      in Munich and Pavia. 86
   Links with physicists at the University? 87
Social relations in Pavia: Ernestina Marangoni and her uncle Carlo,
   a physicist 89
   Ernestina Marangoni. 89
   Carlo Marangoni, a renowned physicist and teacher 90
   Carlo Marangoni: a direct influence on the young Albert? 92

5. Albert's scientific environment in Milan 95
The library of the Lombard Institute (1899-1901). 96
Michele's work on wireless telegraphy 99
   The discovery of electromagnetic waves. 99

Giuseppe Jung's library 101
A thesis by Michele in 1900-1901? 102
An example of following up on scientific questions: thermoelectricity . . . 103
Albert's professional worries 106
Links with his father's new company business 106
Looking for a position as a university assistant 107
Scientific protection 109
*Albert's lasting link with Giuseppe Jung through his personal
library* 109
The professor of chemistry at the Polytechnic Institute 110
The Ansbacher family and the musical environment in Milan 111

**6. Three Albert's scientific questionings (1898-1901) in connection
with his later 1905 work.** **113**
The recurrent problem of the relative motion of matter and ether
(1898-1901) 113
Reading Ernst Mach 113
*An attempt to demonstrate the relative motion of matter
with respect to the ether.* 115
Kyoto 1922 memories 116
The abandoned thesis on molecular forces (October 1900-December 1901) 118
Capillary phenomena and molecular forces in liquids 118
Extending the subject of the thesis to molecular forces in gases . . . 121
Max Reinganum's article in the Lorentz Jubilee volume 122
Abandoning the thesis in February 1902 124
Questions about the nature of light and light quanta from 1901? 125

A final word. **131**

Index of names. **133**

Selected bibliography **135**

Acknowledgements **141**

Introduction

> "It's useful to know as much as possible about what predisposed
> Thales, Archimedes and Galileo to make the discoveries
> they – and you too – gifted to thinking humanity."
>
> Michele Besso to Albert Einstein, 1947.

1905, a "miraculous" year for Albert Einstein?

What more can be written today about the life of Albert Einstein (1879-1955), a universal figure who was the subject of so many biographies and articles? Among his famous biographies are those by Rudolf Kayser, his son-in-law, who published under the pseudonym of Anton Reiser in 1931; Philipp Frank in 1947, his successor at Prague University in 1912; Carl Seelig, who in 1952 retraced his early years in Switzerland, based on eyewitness accounts; Lewis Pyenson in 1985, who provides information on the young Einstein's electrotechnical environment; Abraham Pais in 1982, his colleague at Princeton, who wrote a very well-documented scientific biography; and, in 1997, the excellent summary by Albrecht Fölsing. But this abundance of material does not lift the veil surrounding the great scientist's early life.

It is most surprising that in 1905 Albert, aged twenty-six and a third-class technical expert at the Patent and Intellectual Property Office in Berne, apparently on the fringes of academia, published five major physics papers in the space of a few months, calling into question the very foundations of the discipline. In March 1905, he reintroduced a Newtonian corpuscular nature to light, in what he immediately described as the "revolutionary" form of independent light energy quanta (the future grains of light or photons), whereas previously the entire 19th Century had seen the triumph of a wave theory of light consisting of the vibration of a hypothetical medium, the ether. At the end of April, that year he defended his thesis on the determination of the size of molecules. He then went on to explain the erratic movement of particles suspended in a solution (Brownian motion) and deduced the value of

Avogadro's number, at the heart of the atomic theory of matter. In June, he submitted his paper on the electrodynamics of moving bodies, the foundation of the theory of relativity (restricted to uniform motions); by establishing the invariance of the speed c of light as a postulate, he showed that measurements of duration and length, previously considered absolute, depended on the relative state of motion of the observers. Finally, in September, in a four-page paper, he completed his theory by establishing an equivalence relationship $\Delta E = \Delta mc^2$ between variations in the mass and energy of a body, based on the example of the emission of packets of light waves.

The year 1905 thus appears to have been a "miraculous year", to use the title of John Stachel's book, which presents Einstein's five publications on their centenary in 2005[1]. Yet the notion of a miracle is as foreign to the scope of the scientist as it is to that of the historian of science, and its use also reflects the absence of precise elements enabling us to appreciate the genesis of Albert's ideas. Yet those ideas that took shape so admirably in 1905 certainly had a long history, as the few vague reminiscences Einstein left suggest. His thoughts on relativity date back to 1898, according to his discussions with the German psychologist Max Wertheimer in 1916[2]. However, as his correspondence with his future wife Mileva Marić shows, all he was concerned with at the time was detecting the "relative motion of matter with respect to the luminiferous ether". In December 1901, he was still preparing to write a "capital paper" on the subject and was planning to conduct experiments in parallel. However, special relativity was born in 1905 as a result of the impossibility of detecting this motion using optical or electromagnetic experiments (the absence of a privileged reference frame for describing the laws of physics) and the fact that this observation was established as a postulate. The subject of the size of molecules also had a paradoxical history. The 1905 thesis on the determination of the size of molecules took the opposite direction from an initial thesis begun in October 1900 and abandoned in February 1902. As we shall see, it had focused on mid-April 1901 on molecular forces in weakly compressed gases, with the hypothesis that the actual size of molecules played no role.

As for light quanta, their beginnings seem to date back to April 1901, following Albert's doubts ("misgivings of a fundamental nature") regarding Max Planck's "considerations on the nature of radiation" and with the idea of a possible kinetic character for light in mind. In October 1952, in talks with the physicist Robert Shankland[3], Einstein said that it was a matter "to explain Planck's quantum in more specific terms". To interpret his law of

[1]   John Stachel, *Einstein's Miraculous Year: Five Papers That Changed the Face of Physics*, Princeton, Princeton University Press, 2005.
[2]   Max Wertheimer, *Productive Thinking*, New York and London, Harper and Brothers, 1945.
[3]   Robert Shankland, "Conversations with Albert Einstein", *American Journal of Physics* **31**, 47-57 (1963), here p. 56.

radiation, Planck had introduced a discretisation of the energy of radiating dipoles (oscillators) into elements $\varepsilon = h\nu$, where h is Planck's constant and ν is the frequency of the radiation. Albert Einstein regarded these energy elements as independent entities moving at the speed of light. During these interviews, he also said that the explanation for Brownian motion came to him suddenly. But we shall see that his knowledge of the physical phenomenon itself may date back to the years 1895 to 1900 in Pavia.

The aim of our account is to place the scientific questions that led to the discoveries of 1905 in the general context of the young Albert's years of study: from 1895, the year of his first attempt to enter ETH[4] at the age of sixteen, to 1901, just before he abandoned his first thesis. The information we need comes from Italy, first from Pavia, where he lived, but also from Milan, where he spent the holidays with his family and his friend Michele Besso, who were involved in electrical development. So, it is to the Italian industrial and scientific context of the late 19th Century that we will turn.

Two important witnesses: Mileva Marić and Michele Besso

The main difficulty facing the historian today is the scarcity of available documents, direct testimonies of that bygone era, from which only scattered information remains at first sight, which can only acquire meaning once placed in a general framework. Apart from a few statements made by Albert himself, most of this information can be found in his letters to Mileva Marić, whom he married in January 1903 in Bern, and in his later correspondence with his friend Michele Besso, through their recollections.

Letters to Mileva

Albert Einstein's letters to Mileva Marić were published by John Stachel in the first volume of the *Collected Papers of Albert Einstein* [CPAE][5]. They were edited, with annotations, as a separate work, *Love Letters*, by

---

[4] In what follows, we will use the acronym ETH to refer to the Eidgenössische Technische Hochschule Zurich, or the Swiss Federal Institute of Technology Zurich, which since 1911 has been the name of the Polytechnikum (Polytechnic Institute) of Mileva and Albert's time.

[5] *CPAE (English translation supplement)*, vol. 1: *The Early Years, 1879-1902*, translated by Anna Beck, Princeton, Princeton University Press, 1987. *The Collected Papers of Albert Einstein* [CPAE] is a scholarly, annotated, print and electronic edition of the writings and correspondence of Albert Einstein (1879-1955). Selected from among 80,000 documents, this edition aims at making these writings widely available, for the first time.

Jürgen Renn and Robert Schulmann in 1992[6]. Mileva Marić (1875-1948) was Albert's fellow student in Section VI A of the ETH, which trained secondary school teachers in mathematics and the physical sciences and prepared them for research and an academic career. Their first-year class in 1896 was made up of around twenty students, out of the school's eight hundred divided into seven main sections. Mileva, Albert's elder by just over three years, was therefore twenty-one in 1896.

Born in Titel, a small Serbian town in the autonomous province of Vojvodina, at the time part of the Austro-Hungarian Empire, she was the daughter of Miloš Marić, a soldier in the Austrian army, who in 1867 married Marija Ružić, the daughter of a local nobleman who owned "one of the most beautiful and richest houses in Titel[7]". Mileva's father later bought a farm of two hundred hectares in his native Kać, where the family settled. Mileva had a brother, Miloš, nine years her junior, and a sister, Zorka (1883-1938). Mileva was one of the few girls to study science at the time. Her father – the family being fairly affluent – sent her first to the Royal Serbian High School in Šabac in 1890, where she was taught in German and French, and as a teenager, was allowed to attend an all boys' High School in Zagreb and finally went to Switzerland, where she passed her Matura (Swiss baccalaureate) in Berne in the spring of 1896 before entering the Federal Medical School. Mileva joined the ETH in the second half of 1896-1897. In July 1900, Albert's friends Jakob Ehrat, Marcel Grossmann and Louis Kollros graduated in mathematics. Albert passed, but Mileva failed, probably because of a mark of 2.5/6 in function theory[8]. She repeated the year and enrolled at the ETH for the second semester of 1900-1901, again without success. She had also known since May that she was pregnant, much to the dismay of their respective families.

Albert and Mileva had been seeing each other since the end of 1897 and their relationship became more intimate in 1900. When he was far from her, on holiday with his parents and his sister Maja, Albert wrote to her. To tell her about his parents' opposition to their marriage, or about their holiday plans, his excursions, or simply to give her news of everyday life and talk about their shared acquaintances, but also to share his thoughts on science and his current reading. The letters he sent her between 1898 and 1901 were

---

[6] *Albert Einstein, Mileva Marić: The Love Letters*, edited by Jürgen Renn and Robert Schulmann, Princeton, Princeton University Press, 1992.

[7] Desanka Trbuhović-Gjurić, *Mileva Einstein, une vie*, translation by Nicole Casanova, Paris, Des Femmes, 1991 (French edition).

[8] Mileva's marks in physics, astronomy and her *Diplomarbeit* (an experimental work marking the end of her studies) were only half a point lower than Albert's, and Mileva achieved an overall average of 4/6. Albert, who has 5.5/6 in function theory, has an overall average of around 5/6 (online archive, ETH).

sprinkled with scientific questions, references to physics textbooks and articles. His thoughts are summed up in a few sentences that are quite cryptic for the uninitiated reader, but which nevertheless provide valuable information when grouped together and placed in their scientific context.

Correspondence with Michele Besso

If the letters to Mileva are the main contemporary record of Albert Einstein's scientific questionings, the correspondence with his friend Michele Besso, edited by Pierre Speziali in 1972, *Albert Einstein – Michele Besso (1903-1955)*, is also invaluable[9]. Michele Besso, born in Zurich in 1873 and six years older than Albert, was not only his best friend but also, as he reminded him in his letters, the privileged witness of his scientific interrogations. Their exchanges began at their first meeting, which took place shortly after Michele left ETH and at a time when Albert was applying for admission. An informal collaboration subsequently developed in Milan, where Michele worked between 1899 and 1901, when Albert joined his family during the six-month breaks, and they talked daily, leading him to write from Milan to Mileva in September 1900: "I am spending many evenings here at Michele's". The scientific interaction and connivance of the youthful years turned into a lifelong friendship; Michele was to look after Mileva Marić and the couple's children, Hans Albert (1904-1973) and Eduard (1910-1965), when Albert separated from Mileva to remarry his cousin Elsa (1876-1936) in 1919.

In 1904 Michele joined Albert at the Patent Office in Berne. Carl Seelig explains: "On Einstein's suggestion, Besso had applied in 1904 for a post as examining official at the Swiss Patent Office and had obtained the job"[10]. Michele shared daily discussions with Albert until 1909, when Albert took up a post as professor of theoretical physics at the University of Zurich. There is therefore no trace of letters between 1904 and 1908 in their correspondence. Of the two hundred and fifteen letters, the earliest date from 1903. The February 1903 letter, in particular, illustrates the meticulous bibliographical work that Michele carried out on behalf of his friend, providing him with a complete bibliographical summary of molecular dissociation, one of the subjects of their discussions in Milan in 1901. It is therefore through Michele's later recollections, in his correspondence with Albert, that we must search for traces of their informal collaboration on the most important subjects of the time: quantum physics and relativity.

---

[9]   *Einstein Albert, Besso Michele, Correspondance (1903-1955)*, French translation from German, notes and introduction by Pierre Speziali, Paris, Hermann, 1979.

[10]   Carl Seelig, *Albert Einstein und die Schweiz*, Europa Verlag, Zurich-Stuttgart-Vienna, 1952, London, Staples Press, 1956, p. 71.

The best-known reference to Michele's help is made by Albert in his June 1905 article "On the electrodynamics of moving bodies", in which he lays the foundations of the theory of special relativity: "In conclusion, let me note that my friend and colleague M. Besso steadfastly stood by me in my work on the problem here discussed and that I am indebted to him for many a valuable suggestion[11]". The earliest trace of this informal collaboration appears, in a somewhat allusive form, in the conclusion of a letter sent by Michele to Albert in July 1941. Commenting on a passage in Albert's last book, *The world as I see it*, he wrote to him:

> The theory of relativity, the general one and the most general one, and already the restricted theory, are layered achievements of a very powerful nature. *The World as I see it*, p. 241 ... "all this can only be truly known by the very person who has experienced it". As I write this, I actually have the feeling again of having fallen back into childhood... 1898, and the beginning of 1904, close to you...[12]

So, what does Michele mean by this? What do the years 1898 and 1904 correspond to? The year 1898 refers, as Michele would tell us in 1947, to their reading of Ernst Mach, which he associated with the beginnings of the theory of relativity. The year 1904, on the other hand, refers to quanta, which appear implicitly at the very end of an article Albert submitted in March on the molecular theory of heat, shortly after Michele's arrival in Bern. This quotation also ties in with Michele's still enigmatic assertion of his own role in the quantum problem, when he reminded Albert on January 17, 1928: "For my part, I was your audience back in 1904 and 1905; in the conception of your papers on the quantum problem I provided you with part of your fame, and therefore made Planck a friend of yours"[13]. According to Carl Seelig, who spoke to each of them when writing Albert's biography in 1952, their "main subject of discussion was the discovery of the light quanta" on their return from their work at the Patent Office. But this does not shed any light on Michele's assertion. We will discuss this particular point in Chapter 6.

Michele Besso himself later used subtle images to describe their informal collaboration. In a letter to his aunt dated June 13, 1913, recently found at

---

[11] "On the electrodynamics of moving bodies" [Annalen der Physik **17**, 891-921 (1905)] in *CPAE (English translation supplement)*, vol. 2, Doc. 23, pp. 140-171.

[12] *Einstein Albert, Besso Michele, Correspondance, op. cit.,* (B. 65) p. 213. Translation from French. Einstein writes at the end of the section *Relativity and the ether* of his book *The World as I see it*: "But the years of anxious searching in the dark, with their intense longing, their alternations of confidence and exhaustion, and the final emergence into light, – only those who have experienced it can understand that".

[13] Letter from Michele Besso in *CPAE (English translation supplement)*, vol. 16, Doc. 132, p. 141.

his uncle Marco Besso's foundation in Rome[14], he compared himself to a "pygmy" standing next to a "giant", but then added: "a clairvoyant pygmy":

> In a few days, I think, I'll see my family again, I'll see my dear uncle in Trieste, and if the doctors allow me (but I have no reason to doubt it), I'll come back here [to Zurich] to witness the struggle of my friend Einstein with the Great Unknown: the work and torment of a giant, which I'm witnessing as a pygmy, but as a clairvoyant pygmy.

Albert, for his part, told Seelig about Michele: "I could not have found a better sounding-board in the whole Europe". Michele also took part with him in the gestation phase of the theory of general relativity to which the above extract refers. However, a joint article of around forty pages, written in 1914 half by Albert and half by Michele, was not published. It dealt with the (still premature) application of Albert's theory and the calculations of his friend Marcel Grossmann in 1913 to the calculation of the advance of Mercury's perihelion. Albert would only explain this rigorously with his theory of general relativity at the end of November 1915.

It should also be noted that although Michele Besso collaborated on Albert's scientific problems, Albert also worked for Michele and urged him to start a thesis in 1900, at the same time as he was starting his own thesis as we shall see. Finally, although Michele's presence was clearly important to Albert, he preferred to remain in the background. He also enjoyed taking an interest in a wide range of subjects, such as medicine, psychology and English literature and had a broader interest in the philosophical, economic and social issues of his time.

Michele's unsuccessful plea for a history of the genesis of Albert's ideas

The need to give a framework to Albert Einstein's early scientific ideas stems mainly from the fact that he himself never gave a precise account of their origin, contenting himself with vague allusions, as has been pointed out. In fact, Albert preferred to concentrate on subjects he felt were topical rather than looking back. This is particularly evident in his reply to a letter from Carl Seelig in 1952[15], where he shows far more interest in telling him about his latest research into the unification of the forces of nature (gravitation and electromagnetism), a subject more on the fringes of the scientific development of the time and which he was unable to bring to fruition,

---

[14] Marco Besso Foundation, Largo di Torre Argentina 11, 00186, Rome. Letter from Michele Besso to his aunt.
[15] http://www.library.ethz.ch/en/Resources/Digital-library/Einstein-Online/ Princeton-1933-1955.

than in describing his memories of his youth and his successes, about which
Seelig contacts him! Michele himself had been urging his friend, in vain, for
at least twenty-five years, to retrace the history of his ideas. As early as
December 1921, he informed Albert of his expectations, developing an entire
argument[16]:

> I was again unable to find the complete collection of all your papers until
> 1919 [...]. It understandably meant particularly much to me, because it
> brought to life other remarks made in addition. But why shouldn't this
> rare collection be made accessible to everyone along with the essentials
> for an insight into the historical development of the problems posed, and
> about the novelty of what you had said for reasons of form or substance,
> perhaps even what you would like to have said differently? Many would
> procure the collection at any price, particularly with such an introduc-
> tion, perhaps provided with minimal footnotes. With relatively slight
> effort (I do not estimate it as that slight, because it would demand ear-
> nest retrospection which you may not be fully accustomed to doing) you
> would produce something of immense material and intellectual worth. In
> the introduction I imagine you would convey which works were already
> familiar to you at the beginning of your work: Boltzmann, Lorentz (pre-
> cisely *which* of their writings), then Planck (precisely *which* writings).
> About the outdated papers you would say how, why, by whom; and also,
> what they contain that has not yet been superseded still not fully devel-
> oped. I would not deprive the readers of the published dead ends,[17] they
> are being particularly instructive. What you still regard as valid; what is
> canonical secure and final would be apparent within this context.[18]

The following year, Albert was awarded the Nobel Prize in Physics for his
interpretation of the photoelectric effect based on the theory of light quanta.
By this time, he had become a household name, generating a great deal of
excitement: his theory of general relativity had just been successfully tested
in 1919, giving him the status of a modern-day Newton in the media. A feature of his 1905 publications was the virtual absence of references to earlier
work, so Michele's request was to provide a framework for his research,
so that anyone interested could follow the evolution of his thinking from
the first ideas to their conclusion. Michele probably inherited his historical
and epistemological sensitivity from his uncle Beniamino Besso, as we shall

---

[16] The need for Michele's argument can be better understood by looking at the end
of the letter: "Please don't immediately be negatively disposed. I'm already so enthusiastic about it, quite egotistically, from the outset!". Michele had already been aware
of Albert's lack of interest in the matter for some time.

[17] "I wouldn't hide from the reader the bad leads taken", according to Speziali's
French translation of the German text.

[18] Letter from Michele Besso in *CPAE (English translation supplement)*, vol. 12,
Doc. 338, p. 206.

see later. In all likelihood, he did not receive any response from Albert. Eighteen years later, in 1939, he ventured into this territory himself, writing to Albert on 16 February about a radio script he was to submit:

> This short draft, part of the radio programme to be broadcast here, the rest of the content of which (so they say) would have been submitted to you and would have received your approval, should have been in your hands a long time ago; you could then have written to me to touch it up or add to it. I would at least have sent it to you! The way you've managed to wake up one Sleeping Beauty after another is like something out of a fairy tale. Michelson's experiment with Fresnel's drag coefficient, Riemann's hypothesis about the shape of space according to its content, statistical mechanics and Planck's quanta, which were still very young and had already been forgotten in 1904 [...][19].

But the following week he wrote: "The text of the conference I sent you will not be used. It has been refused, with good reason, I believe, because the names of the people mentioned mean nothing to the listeners". Clearly, the journalists, focused on the personality of Albert, whose communication in the media appeared to be highly regulated, ignored the themes dear to his friend. In October 1945, as he grew older and felt that there was not much time left to faithfully retrace the history of Albert's ideas, Michele once again returned to the subject with insistence: "A precise exposé, with your 'bad leads' and your 'buried hopes', would be of immense value for the evolution of science. You certainly have an assistant who is interested in the history of science and could collaborate on it[20]". The terms "bad leads", "buried hopes" and the inverted commas placed by Michele suggest that these are the very terms used between them to describe the trial and error during the building of theories. In December 1947, a tension even arose for the first time when Michele spoke in the media about Mach's role in relativity, as we shall see in Chapter 6. This would be his last known attempt, at the age of seventy-five, to encourage Albert to explain the history of his ideas. It is from this letter that we extract the sentence at the beginning of the chapter: "It's useful to know as much as possible about what predisposed Thales, Archimedes and Galileo to make the discoveries they – and you too – gifted to thinking humanity."

---

[19] *Einstein Albert, Besso Michele, Correspondance (1903-1955), op. cit.*, (B. 59) p. 199. Translation from French.
[20] *Einstein Albert, Besso Michele, Correspondance (1903-1955), op. cit.*, (B. 70) p. 220. Translation from French.

When, five years later, Albert received a letter from Carl Seelig on February 25, 1952, announcing his intention to write a biography of his Swiss period, he took the initiative and sent a series of instructions to Michele on March 6:

> I already know that the good Seelig is looking after my childhood. It has to be said that we discussed scientific matters every day on our way home from work. I should also mention my friendship with Maurice Solovine (currently in Paris) and with the two brothers Conrad and Paul Habicht, whom I met when I was working as a private teacher in Schaffhausen. Paul H. died recently. He built a small electrostatic field machine for me, which I used for static measurement of low voltages by multiplication. In Bern, I regularly had evenings of philosophical reading and discussion with C. Habicht and Solovine, during which we dealt mainly with Hume (in a very good German edition). This reading had a certain influence on my development alongside Poincaré and Mach. It was you who recommended the latter to me during my student years, after we had met at Madame Caprotti's house[21].

While the development of Albert Einstein's ideas prior to 1905 is a particularly difficult task to reconstruct in any detail today, it is nonetheless a fascinating and necessary undertaking, according to Michele himself. Michele's testimony in the form of as yet unexploited family correspondence, or perhaps his memoirs (if he wrote them), would be a major discovery for understanding the genesis of Albert Einstein's scientific ideas, which revolutionised physics. Given the incomplete nature of the evidence at our disposal, we have no choice but to follow in the footsteps of the young Albert, more than a century apart and attempt to fill in the gaps left by historiography on the years during which his scientific orientations were determined. The approach is akin to a police investigation, following in the footsteps of missing witnesses. An investigation in the field, in the places that the players visited regularly. In institutional archives and family memories too. The aim is to draw the contours of a scene in which these actors can be positioned and then to superimpose the young Albert, in touch with them and the reality of his time.

Editorial plan

Our starting point will be the Einstein family's move to Italy. Albert's uncle, Jakob Einstein (1850-1912), was an engineer who graduated from the Stuttgart Polytechnic. He persuaded his elder brother, Hermann (1847-1902), Albert's father, to join him in 1880 in Munich in a first company supplying

---

²¹ *Einstein Albert, Besso Michele, Correspondance (1903-1955), op. cit.*, (E. 88) p. 272. Translation from French.

water pumping and gas installations and then to embark on the adventure of industrial and domestic electricity, which was beginning to develop. Following business difficulties in Munich, where they had employed up to two hundred people, in 1894 they transferred their business to Milan and Pavia, where they set up their new company, the Einstein-Garrone National Electrotechnical Workshops. After the sale of the company in September 1896, the family stayed in Milan until Hermann's death in October 1902.

In order to understand the environment in which the Einsteins' company was established in Lombardy, we need to go back over the stages of Italian unification in Chapter 1, entitled "The Italian political and industrial context in the 19th Century". It was during the preceding wars of independence that the links between the political, economic, industrial, scientific and literary leaders who were working to modernise the country were forged. The earlier influence of the Napoleonic era is also recalled in connection with the creation and role of the Lombard Institute, the academy of science and letters in Milan, where Albert worked on his degree and dissertation. We will delve into the milieu of the engineers, professors and senators of the Kingdom of Italy, who organised the scientific and industrial landscape, such as Giuseppe Colombo for the electrical sector, or the father of Italian electrical engineering, Galileo Ferraris, figures with whom the Einsteins had close ties. The railway industry will be mentioned in connection with the development of the electrical industry and workers' insurance, where the Besso family will feature prominently. At the same time, engineering schools inspired by the French and German polytechnics came into being with the first reforms of the new state. Lastly, the Hoepli publishing house will provide the opportunity to follow these developments from a different angle.

The 1880s saw the beginnings of electric street lighting, which gradually replaced gas lighting in towns and cities. A new community of electrotechnical engineers was formed in Europe, with its own journals. The proliferation of international exhibitions on the subject of electricity, the first of which was held in Paris in 1881, will be our starting point for defining the electrical context of the Einstein company more precisely in Chapter 2 entitled "The company of Jakob and Hermann Einstein in the international electrotechnical context – exhibitions and reviews".

We will supplement the presentation of the Paris Exhibition with that of 1884 in Turin for alternating current and the presentation of the Gaulard-Gibbs electric transformer studied by Galileo Ferraris. These two exhibitions clearly played an important role in the Einsteins' entrepreneurial orientation, since their first electrotechnical activity appeared at the 1882 Munich Exhibition and they specialised and developed their company in 1885, while turning their attention to Italy. We will also look in detail at their participation in the 1891 Exhibition in Frankfurt, where Heinrich Friedrich Weber, Albert's professor of physics and electrical engineering at the ETH in Zurich, headed the commission examining the first three-phase alternating current power line.

In Chapter 3, entitled "Michele Besso and the role of his family in Italian industrialisation", we will look at Michele's personality and the circumstances – still unclear – of his meeting with Albert. The links between the two friends require us to take a closer look at Michele's family. His father Giuseppe and his uncle Marco were in charge of the General Insurance company (*Assicurazioni Generali,* today *Generali*); his uncle Beniamino was in charge of the railways and the author of a number of works for the dissemination of science; and his youngest uncle Davide was a university mathematics professor and founder of an educational periodical. On the Cantoni side, Michele Besso's maternal branch, we will meet Vittorio Cantoni, an uncle who graduated from ETH in Zurich and Polytechnic in Milan, and who installed Italy's first alternating current distribution network in Tivoli, and Giuseppe Jung, a member of the Lombard Institute and professor at Polytechnic, an uncle by marriage to Michele to whom Albert sent his articles when he was looking for an assistant post. We will also look at the influence these uncles may have had on Michele's personal development.

Chapter 4, "Albert's environment in Pavia and preparations for ETH", looks at his scientific environment in 1895. When Albert left Munich in December 1894 to join his family in Milan, he did not have his baccalaureate and was not at school in Italy. Nor was he old enough to sit the entrance exam. But that doesn't stop him from considering a place at one of Europe's most prestigious schools, ETH Zurich. We will relate his first scientific paper, written in the summer of 1895 and entitled "On the investigation of the state of the ether in a magnetic field", to this first attempt. As he did not work entirely alone in preparing it, we will describe the scientific environment of his uncle's engineering office and show that the company's main partners, Lorenzo Garrone and Angelo Cerri, had links with the University of Pavia. Finally, we will look at the family of Ernestina Marangoni, the mutual friend of Albert and his younger sister Maja, drawing attention to her uncle, the physicist Carlo Marangoni, whom Albert met in September at Casteggio between 1895 and 1900 and who may have directed his early interests towards capillarity and perhaps even Brownian motion.

The starting point for our study in Chapter 5, entitled "Albert's scientific environment in Milan", will be the place from which the letters to Mileva were written: Milan, which until now had not been associated with any of Albert's scientific activities. After completing his secondary education in Aarau, Switzerland, with Jost Winteler's family in 1895-1896, Albert went on to study at the ETH between 1896 and 1900, and then as a thesis student in 1900-1901. He spent two to three months a year in Milan during the spring and summer half-term holidays, as well as at Christmas, where he devoted himself to his bibliographical research. Based on information taken from his letters to Mileva, we will establish that he studied at the library of the Lombard Institute in Milan between 1899 and 1901. Based on his critical reading of articles and works, we will follow the development of his scientific questionings on his first subject, thermoelectricity, the subject of his final dissertation with Mileva. We will also look at wireless telegraphy, on which Albert worked for Michele, in relation to the contents of his uncle Giuseppe Jung's library. We will then return to the personalities who provide him with "quite good connections", as he informed Mileva: Giuseppe Jung and Bernardo Ansbacher, who probably put him in touch with the chemistry teacher Luigi Gabba.

Finally, in Chapter 6, "Three Albert's scientific questionings (1898-1901) in connection with his later 1905 work", after going back over Mach's reading of 1898, we will discuss Albert's attempts to detect the relative motion of matter with respect to the ether, which he recalls in his Kyoto lecture of 1922. We will then discuss the subject of his thesis on molecular forces in liquids and capillarity, which he began on leaving ETH at the end of 1900, then extended to weakly compressed gases in April 1901, probably following his reading of an article by Max Reinganum published in the *Jubilee* for Lorentz and recently registered by the library of the Lombard Institute. Finally, we will clarify the hypothesis proposed by Jürgen Renn in 1993 that Albert may have formulated an idea of light quanta as early as 1901 on the basis of his reading of Planck and a kinematic vision of light, the possible origins of which we will examine.

1

The Italian political and industrial context of the 19th Century

We begin our account by placing the Einstein family's electrotechnical activities, which began in Munich in the early 1880s and were later transferred to Lombardy, in the historical, political and industrial context of nineteenth-century Italy. Albert's family (his father Hermann, his mother Pauline Koch and his younger sister Maja) and the family of his uncle Jakob Einstein moved to Milan in the summer of 1894. Albert, who had initially stayed in Munich to complete his studies at the Luitpold Gymnasium, left the school in December to join his family unexpectedly at Christmas. In one of his letters to Carl Seelig in 1952, he summed up the situation in Italy by saying: "My father, together with my mother, lived partly in Milan and partly in Pavia [the factory's production site]". Hermann and Jakob were given the opportunity to play an even more active role in the electrification of a country to which they had been contributing since at least 1887. They brought their technology and know-how to the table and became part of an industrial fabric that was undergoing major restructuring. Their adventure thus became part of the industrial history of Italy, a country that had not yet been thirty years old since its birth in 1861 and still a territory to be conquered.

The houses in which the Einsteins live, previously occupied by patriots who played a leading role in unification, may have been made available to them, placing our story in a historical and political perspective. Indeed, the influence of the revolutionary period of the *Risorgimento*, which led to unification, is still very much alive at the end of the 19th Century, as the Italian ruling class was forged within this common ideal during the wars of independence from 1848 to 1866. We will give an overview of this period, going back to the Napoleonic era for the creation and role of the Lombard Institute, Milan's academy of science and letters, which links the academic, engineering and industrial worlds. At the end of the 19th Century, Italy was rapidly catching up with the other major European countries in terms of industrial development and was turning to the northern countries of the peninsula for its development, financed by banks and insurance companies. Engineering schools, created in the wake of unification and university reform, adapted to the needs of the railway and electricity sectors. We will also follow this development through the Hoepli publishing house and scientific book-shop in Milan.

The Einsteins in Milan and Pavia: settlements and relationships

The short adventure of Einstein, Garrone & Cie (1894-1896)

The Einsteins' new company was founded in Pavia on March 14, 1894 at the office of notary Davide Giulietti (on Albert's fifteenth birthday)[22]. The town council approved the business plan in June, and the factory prem-ises were delivered at the end of September. Jakob and Hermann Einstein joined forces with engineers Lorenzo Garrone and Angelo Cerri to found the Einstein, Garrone & Cie National Electrotechnical Workshops, which they registered with the Pavia Chamber of Commerce. Its capital of 105 000 lire was divided between the Einstein brothers (30 000 lire) and engineer Lorenzo Garrone (30 000), engineer Angelo Cerri (15 000) and other partners. A note dated October 25, 1894, from Davide Giulietti states that the company's cap-ital was increased by a loan of 50 000 lire from engineer Giovanni Barberis, who was to become Michele Besso's manager in Milan.

The company was involved in the construction of DC and AC dynamos and electric motors, power transmission, arc lamps and electrical appliances, an activity that we will consider in the context of electrical development in the next chapter. The company employed around eighty people and the headquarters were established in Milan, first at 2 via Berchet and then at 41 via Manzoni, in the Palazzo Borromeo d'Adda. It also had a subsidiary in Turin, at 6 Corso Duca di Genova (now Corso Stati Uniti). The list of customers in the company's advertising catalogue includes: 22 municipali-ties, 19 of which are in Italy, 2 in Germany and 1 in France; 100 companies, 71 of which are in Italy, 27 in Germany and 2 in France. Two-thirds of the towns and half the companies are in Piedmont. In Piedmont, Lombardy and Liguria, in particular, the Einsteins' company provided electrical lighting for the towns of Susa, Bardonecchia, Perosa Argentina, Agliè, Coggiola (Biella), Varallo Sesia, Sannazzaro de' Burgondi, Varzi (Voghera), Ovada, Pieve,

---

[22] The details of the Einstein company in this paragraph are taken mainly from the book by Sergio Biscossa, *Gli Einstein imprenditori a Pavia (1894-1896)*, Mortara, Rotary Club di Vigevano, 2005.

Pieve di Teco and Cogoleto[23]. In its catalogue, it highlights the installation of an electric dynamo in the ducal castle of Agliè, the residence of "his royal highness the Duke of Genoa", married to Isabelle of Bavaria and brother of Victor-Emmanuel II, King of Italy, attesting to the Einsteins' connections at the highest political level in Italy.

The Einstein, Garrone & Cie enters into a partnership in Pavia with the engineer Losio, acquiring shares in the Alessandro Volta company for the city's electric lighting. Losio had to ratify the transfer of a concession to harness the water of the Ticino river to produce hydroelectric power. The transfer of these rights was part of the cause of the Einsteins' difficulties. The Alessandro Volta company sought the intervention of Giuseppe Colombo, a leading figure in the Italian electricity sector, who entrusted the management of the business to Luigi Zunini, director of the Carlo Erba Institute of Electrical Engineering at Milan Polytechnic and a professor at the school. This decision led to the liquidation of the company Einstein, Garrone & Cie on June 1, 1896 and its sale by auction on August 26. On September 24, 1896, the company was bought by Carlo Monti & Cie, based in via Manzoni 41, Milan. The company itself was the result of a capital increase by C. Monti & Cie, founded on May 5, 1895 by Carlo Monti, who had already acquired shares in Einstein, Garrone & Cie. Giuseppe Colombo was one of the three people who oversaw the acquisition of the Einsteins' company, along with Milan lawyer Aureliano Scrosati and engineer Giovanni Battista Pirelli[24]. Carlo Monti was one of Colombo's close collaborators: he had just graduated from the Milan Polytechnic in 1882, where he had been taught by him, and had worked with him on the lighting of the La Scala Theatre the following year.

Famous homes and political connections

The Einsteins lived in a number of famous homes in Pavia and Milan, which are worth mentioning because they provide an insight into their political and social environment. When they arrived in Milan in the summer of 1894, they initially lived in Palazzo Ricordi, 2 via Berchet, which was also their company headquarters, a short walk from the cathedral square. In Pavia, they lived in via Foscolo 11, in the house that had been rented by the famous Italian poet Ugo Foscolo in 1807-1808, when he held the chair of eloquence at the university. Albert's application for admission to Zurich Polytechnic in October 1896 bears this address. It was a beautiful fourteenth-century

---

[23] Renato Giannetti, "Tecnologia ed economia del sistema elettrico", in Giorgio Mori (Ed.), *Storia dell'industria elettrica in Italia*, vol. 1, *Le origini, 1882-1914*, Bari, Laterza, 1992, p. 375.

[24] Document from the Milan Chamber of Commerce concerning Carlo Monti's company. Pirelli, a graduate of Milan Polytechnic, founded Italy's first tyre industry in 1872 and launched the manufacture of insulated electric cables in 1880.

mansion, Palazzo Cornazzani, with an inner courtyard decorated with frescoes painted on the walls of the rooms. Ernestina Marangoni, a friend of Albert and his sister Maja in Pavia, recalled in 1955 in her column in the newspaper *Provincia Pavese* in tribute to Albert Einstein, that the family had a living room, a reception room and a music room. The property was purchased in 1873 by Luigi Doglia and inherited by his sons Roberto and Goffredo in 1878, as documented in the State Archives in Pavia. The house appears to have been made available to the Einstein family or their associates, as there is no record of a lease in their name[25]. The presence of the Einsteins on this property is probably linked to the hydraulic sector, in connection with the concession to exploit water from the Ticino mentioned above, since Luigi Doglia had been president of the Verrua hydraulic consortium on the Po river in his time, as stated in the 1874 guide-almanac for the province of Pavia; his son Roberto, who appears to have been the sole owner at the time of the Einsteins, was a lawyer in Pavia. Following the sale of their business, the Einsteins settled permanently in Milan in October 1896, at 21 via Bigli, a stone's throw from the La Scala Theatre, in a flat that had been occupied by Countess Clara Maffei (1814-1886). The building, purchased in 1879 from Gian Giacomo Poldi Pezzoli, founder of Milan's famous museum of the same name, was owned by the young prince Luigi Alberico Trivulzio. The Einsteins' flat consisted of "eleven bedrooms with two sumptuous reception rooms, connected by a large archway[26]". This seems an astonishing level of luxury, in stark contrast to Maja Einstein's claim that, at the time, "the family had hardly anything left[27]" following their industrial misfortunes.

The places inhabited by the Einstein family in Pavia and Milan are no strangers to the world of the *Risorgimento* that led to the unification of the country. Clara Maffei was a famous figure in Milan's political and intellectual life. After separating from her husband, the writer and poet Andrea Maffei (1798-1885), she lived at via Bigli 21 from 1849 until her death in 1886. Here she kept a salon regularly by the reformist and patriotic political elite of Lombardy; her closest friends were the famous composer Giuseppe Verdi and the writer Carlo Tenca. The mathematician Francesco Brioschi, founder of the Milan Polytechnic, frequented her salon, as did Giuseppe Colombo, who organised the Italian electrical industry[28] and succeeded Brioschi as director of the school. In a letter to Clara Maffei in October 1882, Carlo Tenca told

---

[25] Lucio Fregonese, *Gioventù felice in terra pavese. Le lettere di Albert Einstein al Museo per la Storia dell'Università di Pavia*, Milan, Cisalpino Istituto Editoriale Universitario, 2005. Translation from Italian.

[26] Fabio Bevilacqua and Jürgen Renn (Eds.), *Albert Einstein, ingegnere dell'universo*, Milan, Skira, 2005, p. 43.

[27] "Albert Einstein - A Biographical Sketch" (Translated Excerpts) by Maja Winteler-Einstein in *CPAE (English translation supplement)*, vol. 1, pp. xv-xxii.

[28] Rita Cambria, "Giuseppe Colombo", in *Dizionario biografico degli Italiani*, vol. 27, 1982.

her of his enthusiasm for the creation of Colombo's first electric company, whose home would be the first to be lit in Milan. It should be noted that the Italian ruling class of the 1890s was made up of men who, thirty years earlier, had taken up arms for Garibaldi, like Brioschi and Colombo, and who were attracted by the revolutionary ideas of Mazzini.

The fact that the Einsteins had links with leading Italian political figures outside – or connected with – Colombo is confirmed by a recollection of Maja Einstein, who wrote in her memoirs: "The hot summer of 1895 was spent in Airolo on the Gotthard, where young Albert gained a fatherly friend in the Italian minister Luzzatti, who happened to be staying there[29]". Luigi Luzzatti, then Finance Minister, had just succeeded Giuseppe Colombo as Finance Minister for the second time. Both were professors at Milan's Polytechnic and had been called upon to set it up by Francesco Brioschi. Maja's statement also opens up a new perspective. The fact that the Einsteins, industrialists in the electrical sector, met Luzzatti at Airolo, the Swiss-Italian end of the Gotthard tunnel, certainly indicates links with the railway sector. At the end of the 1880s, their company had already lit the Bardonecchia railway station at the Italian end of the Fréjus tunnel. A few years later, Maja married Paul Winteler, the last son of Jost Winteler, with whom Albert had lived in Aarau from 1895 to 1896 when he was preparing for his Matura. Pierre Speziali explains that "a lawyer by training, Paul Winteler had joined the management of the Gotthard railways". We will see throughout our account that there are real links between the rail and electricity sectors. But first, let's take a look at the unification of Italy in the 19th Century to see how this second phase of industrialisation, in which the Einsteins were to participate, took shape.

Historical context: Napoleonic and *Risorgimento* influences

The Cisalpine Republic

Following his victory over the Austrians, Napoleon Bonaparte proclaimed the Cisalpine Republic in June 1797. One result of the unification of the territories of the Cispadane Republic (south of the Po) with those of the Transpadane Republic (mainly Lombardy) in the north of the peninsula, was that the new Republic chose Milan as its capital. Its operation and administration were inspired by the French model. Other "sister republics" were proclaimed, such as Genoa. Deposed by the Austro-Russian armies in 1799, they were restored by Napoleon in 1801. The Cisalpine Republic was reborn

---

[29] "Albert Einstein - A Biographical Sketch" (Translated Excerpts) by Maja Winteler-Einstein in *CPAE (English translation supplement)*, vol. 1, p. xxii.

and expanded. In 1805, it became the Kingdom of Italy, with Eugène de Beauharnais as Viceroy. The Lombardy bourgeoisie benefited from this imperial period. It developed profitable crops such as rice, corn and mulberry trees for rearing silkworms used in the textile industry. Its share of land increased from 24% to 40% between 1789 and 1804. It was hard hit by the restoration of the old aristocratic regimes by the Austrians after Napoleon's exile and remained fundamentally opposed to the re-establishment of the old borders of the duchies, which entailed the payment of customs duties for the transit of goods and hindered the movement of people and ideas. The country aspired to reunification, whether on the basis of firm opposition to the aristocracy or the clergy with revolutionaries inspired by the French Revolution, or on the basis of an economic model favourable to the bourgeoisie.

Napoleon was as popular as he was rejected. The Milan City Museum has a large number of period documents and paintings illustrating the different phases of his adventure, bearing witness to the imprint he left on the Lombardy ruling circles. Napoleon modernised the institutional framework of the Italian republics based on the French model, for example with the 1808 decree on higher technical education, which introduced university courses and diplomas for engineers. But it also disappointed the aspirations for freedom that it had given rise to. The example of Ugo Foscolo is revealing: the famous poet, who had ardently supported Napoleon's enterprise at the outset, later turned his back on it, believing that Napoleon had betrayed the people's aspirations for unification by ceding the Veneto to the Austrians. As a committed poet, he inspired young people to aspire to unification, a desire to which the revolutionary Giuseppe Mazzini himself would have been sensitive. It should be noted that, in parallel with the spread of the ideas of the French Revolution, Masonic lodges developed in Italy, which were renewed during periods of insurrection. So, it's not surprising to find there some of the great names of the period, such as Eugène de Beauharnais, Ugo Foscolo, Giuseppe Mazzini, Giuseppe Garibaldi, etc.[30].

The Austrians returned to Lombardy in 1815 after Napoleon's exile, but the experience of Napoleon's regime, which relied on a local elite, encouraged the Lombards to become more independent. The revolutionary ideal and the Romantic literary movement fuelled the desire for greater freedom, with the liberalisation of the press and an increase in the circulation of magazines and novels. The best-selling novel *The Betrothed* by Alessandro Manzoni (1785-1873), published in 1827, was an indirect call to arms against the Austrians. The most radical opposition to the current regime came from Giuseppe Mazzini (1805-1872), a native of Genoa who, in exile in Marseilles, founded the socialist-inspired Young Italy movement in 1831. He advocated the elimination of the existing states, the establishment of a republic and the

---

[30] Anna Maria Isastia, "La Massoneria", in *L'Unificazione*, 2011, online at www.treccani.it.

end of the Catholic Church, within a federal Italy. However, the many armed insurrectionary movements it unleashed ended in failure and the leaders were executed. Milan's modernist bourgeoisie and aristocracy preferred to choose a less radical path, but still made an effective contribution to the reform movement. They sought the military support of the Savoy family, which ruled the kingdom of Piedmont-Sardinia, the only sovereign and modernist state on the peninsula at the time. The historian Pierre Milza reports that from 1830 onwards, the major scientific congresses "each time brought together hundreds of participants, mostly teachers, engineers, technicians and civil servants, thus creating a milieu of reformers who made a powerful contribution to the formation of a collective national consciousness[31]". Protest movements against Austrian rule developed in Italy at the same time as the abdication of Charles X in France and the advent of Louis-Philippe, who embodied a policy favourable to the industrial revolution.

The insurrection of 1848

A few years later, the Five Days, from March 18-23, 1848, marked the insurrection of the city of Milan against the Austrian government. It was triggered by the French revolution of February 1848 against the constitutional monarchy of Louis-Philippe, which led to the advent of the Second Republic. The Austrian army, led by Marshal Radetzky, retreated to northern Lombardy. The Milan War Council was placed under the command of Carlo Cattaneo, who organised operations from the Taverna Trivulzio palace in via Bigli 9. In his chronicle of the days of 1848, Cattaneo wrote:

> To get there [to the command centre], all we had to do was to cross via Bigli, a narrow, winding street that was very easy to barricade [...] Cernuschi obtained the key to the gate opposite Mr Manzoni's house, had a passage cut in the wall of the Belgiojoso garden and placed sentries on the walls of the other gardens [...] This group of beautiful houses, courtyards and gardens had at its centre a fine 17th Century house, the former Taverna house, occupied by the French consulate, where, alongside our flag, the colours of the French Republic flew. Our faith in the friendship of this great nation was not without influence at this solemn moment, when an entire people was throwing itself, with so few resources, into the bloody path of its regeneration[32]!

The Milan municipality's appeal to Charles Albert, King of Piedmont and Sardinia, to push the Austrians back to their border, marked the start of the First War of Independence. A year later, the war ended in defeat and Charles Albert's abdication in favour of his son Victor Emmanuel. Cattaneo blamed

---

[31] This paragraph and the next two are based on Pierre Milza, *Histoire de l'Italie, des origines à nos jours*, Paris, Fayard/Pluriel, 2013.
[32] Carlo Cattaneo, *The Milan Uprising*, in *Tutte le Opere di Carlo Cattaneo*, IV, Luigi Ambrosoli (Ed.), Verona, Mondadori, 1967, p. 213.

the failure on Charles Albert's political calculations and hesitations. The Austrians' return to Lombardy was followed by a crackdown on the insurgent towns. The leaders of the protest, including Carlo Tenca in Milan, Princess Belgiojoso (Cristina Trivulzio Belgiojoso), Carlo Cattaneo and others, were forced into exile. Princess Belgiojoso paid tribute to the Savoy family in her *History of the House of Savoy,* published while in exile in Paris. In 1852, Cattaneo set up a technical college for modern education in Lugano, where he was joined by the physicist Giovanni Cantoni, the future rector of the University of Pavia, who was also in exile and whom we shall see again in the course of our story. Cattaneo concludes his chronicles of the insurrection by placing the ideal of unification in a European perspective:

> Now, it is by another link that we must think of reuniting the great family of European nations and we must not look for it in the unity of power, but in the freedom and equality of all peoples [...] We shall have peace, and be able to enjoy it, when we have the United States of Europe.

The Risorgimento and the birth of a modern state

The period from 1848 to the unification of Italy in 1861 is known as the *Risorgimento.* Italy was born following the second war of independence led in 1859 by Victor-Emmanuel II, King of Piedmont and Sardinia, and his minister Camillo Cavour, "the prototype of the liberal aristocrat open to modernity but a declared opponent of extremism", as Milza describes him. Cavour had turned to France to finance Piedmont's infrastructure, in particular its railways, making it the most advanced state in the peninsula. The revolutionary Giuseppe Garibaldi (1807-1882), a native of Nice, entered the service of Victor Emmanuel. He disembarked in Sicily with "the Thousand", dressed in the famous red shirts, then made his way up to Naples to hand over the Kingdom of the Two Sicilies to him, but was stopped by the latter in his advance to the gates of Rome. Victor-Emmanuel felt that the political conditions were not right to take Rome, which was protected by France. Umbria and the Marches were conquered by the Piedmontese army, which tacitly guaranteed papal power over the Lazio region and Rome. The Second War of Independence ended with the proclamation of Victor-Emmanuel II as King of Italy in 1861. In exchange for its support, France received the counties of Savoy and Nice, after Tuscany and Emilia-Romagna were attached to the Kingdom of Italy. Veneto became part of Italy in 1866 after the Third War of Independence. It was not until 1870, taking advantage of the weakening of France after its defeat by Prussia, that the Kingdom of Italy extended to Rome, which it took as its capital in 1871. There remained the question of the irredentary lands, such as Trieste, which remained under Austrian domination, and which would not be resolved until Italy entered the war against the Austro-Hungarian Empire in 1915.

From the 1860s onwards, a new era began to take shape, corresponding to the industrialisation of the country and major political and economic changes. At university level, this was reflected in a far-reaching reform of higher education, to which the likes of Francesco Brioschi and Pietro Blaserna contributed. Italy, which had initially relied on textiles to develop its industry, had fallen behind in the more modern sectors. So industrial bosses, who had financed Garibaldi's arms drive, were now investing in this new sector by buying shares in new companies and banks. They also subsidised professorships and institutes at Milan Polytechnic: Eugenio Cantoni, head of Cantoni Spinning Mills, the first Italian industry to be listed on the stock exchange, financed a course in industrial economics in 1871, and Carlo Erba, head of the pharmaceutical industry, founded the eponymous electro-technical institute in 1886.

The creation and role of the Lombard Institute

The link between engineering, academia, finance and politics, necessary for the effective development of industry, was fostered in Milan by the Lombard Institute set up at the beginning of the 19th Century by Napoleon Bonaparte. This is the present-day Lombard Institute, an academy of science and letters modelled on the Institut de France. Its history is described in detail in the three volumes recently published[33]. Originally based in Bologna, it was made up of sixty members, including thirty honorary members, appointed by Napoleon on November 6, 1802. The institute was divided into three classes: physical sciences and mathematics, moral and political sciences, literature and the arts. For the Cisalpine Republic, its purpose was to be a "national institute" responsible for gathering discoveries and improving the arts and sciences. On January 15, 1804, the academicians elected their first president: the physicist Alessandro Volta, who had presented his electric battery to Napoleon at the Academy of Sciences in Paris. Volta's work marked the beginning of the era of electricity, which until then had been confined to the field of static electricity, which was of little use. The institute's focus on developing a scientific and industrial culture led to the creation of a technological cabinet, where Italian and foreign inventions that had won awards from the academy were displayed. In 1810, as King of Italy, Napoleon gave the institute the title of Royal Institute of Sciences, Letters and Arts, and moved it to the Palazzo Brera in Milan. When Napoleon fell, the Austrians reinstated their old administrative system, but kept the institute under the name of Royal Imperial Institute of the Venetian-Lombard Kingdom. The Venetian Institute became independent in 1838. The word arts, reserved for

---

[33] Emilio Gatti and Adele Robbiati Bianchi (eds.), *Istituto lombardo accademia di scienze e lettere*, 3 vols., Milan, Libri Scheiwiller, 2007-2009.

the Academy of Arts in Brera, was removed from the name and the Lombard Institute took on its current name.

A new impetus came in 1859 with the definitive departure of the Austrians. The aforementioned writer Alessandro Manzoni was acclaimed President. One of the major issues facing the young state was the choice of Italian language. Manzoni promoted a language based on Tuscan, of which the poet Dante Alighieri was the emblematic figure. Graziadio Isaia Ascoli (1829-1907) was one of the leading figures to join the Institute at the time. In opposition to Manzoni, he advocated an Italian language based on contributions from the various dialects[34]. An effective member of the Institute in 1864 and of the Accademia dei Lincei in Rome, he was president of the Milan Scientific-Literary Academy, set up in 1859 by the Casati law. This institution anticipated the creation of the Faculty of Letters and Philosophy at the University of Milan in 1924. Ascoli's name comes up again in his correspondence with his nephew Marco Besso, Michele's uncle. Giuseppe Colombo, one of the leading figures in the electrical industry, mentioned in connection with the Einstein company, was twice vice-president (1890-1891; 1894-1895) and President (1892-1893; 1896-1897) of the Lombard Institute at the end of the 19th Century.

The Institute was not the only cultural and intellectual centre in Palazzo Brera. In 1810, Napoleon also founded the famous Pinacoteca di Milano, one of Italy's most important museums. One of the most famous paintings to be found there is *The Kiss* by the painter Francesco Hayez, who supported the cause of unification. The picture was painted in 1859 and depicts a couple kissing, illustrating the union of France and Italy. The statue of Napoleon – which has just been renovated – still stands in the centre of the courtyard of Palazzo Brera. The palace is also home to the Milan Observatory, founded by the Jesuit astronomer Roger Boscovich in 1760, where he taught mathematics. The palace also houses the famous Braidense Library, with its one million books, the oldest of which date back to the 15th Century. The library was created by Maria Theresa of Austria in 1786 on the premises of the Jesuit College, after the Jesuits had been dissolved in 1773 by Pope Clement XIV. Maria Theresa of Austria also enlarged the botanical gardens adjoining the palace and created the Brera Academy (for the fine arts). The Palazzo Brera was therefore undeniably the cultural and intellectual centre of Milan, and experienced intense life in the second half of the 19th Century. Artists and intellectuals were also to be found on the terraces of the many cafés in the area. This atmosphere, steeped in history, can still be felt today,

---

[34] It should also be noted that as early as 1851, Graziadio Ascoli proposed a universal language for telegraphy, pasitelegraphy, an application to the telegraph of pasigraphy (universal writing) and pasilaly (universal language). It is defined as "the art of corresponding from afar by signals with all the peoples of the world"; see Comte de Firmas-Périés, *Pasitélégraphie*, Metzler, Stuttgart, 1811 (note by Alessio Moretti).

as you stroll along the cobbled streets and see the 17th Century houses that adorn this former suburb of old Milan.

This famous place was frequented by Albert Einstein, according to his son-in-law Rudolf Kayser, as we shall see later, which we shall relate to his bibliographical work.

The new industrial and economic framework

Unification and railways

From the 1860s onwards, the railway symbolised the unification of Italy and its industrial development, so much so that one Minister of Labour described it as "the architect of Italian unity", as historian Albert Schram[35] notes. The aim was as much to erase the inequalities between Italy's different regions as to enable the country to integrate more fully with Europe. The railways absorbed half of the State's investment in the decade between 1860 and 1870. The mixed private-public system for operating the lines was not standardised until they were nationalised in 1905. The 1865 law establishing five major companies in Italy was co-signed by Finance Minister Quintino Sella (1824-1885). The largest of these was the Franco-Austrian Upper-Italy Railways company, founded by privatising the Piedmontese Railways company, of which Camillo Cavour, Victor-Emmanuel II's minister, had been the director. It also included the Tuscan network and was headquartered in Milan in 1874. Its network now covered 2 422 km. The nationalisation of the company in 1878 was the largest financial operation ever recorded for a railway company in Europe. The Royal Railway company of Sardinia was one of the other four companies. It was founded in London in 1863 and later managed by Beniamino Besso, an uncle of Michele's who had previously worked for the Upper-Italy Railways company. In 1885, a new structure for the Italian network was put in place and three major companies now managed the rail network: the Sicilian Railway company, the Mediterranean Railway company and the Adriatic Railway company, with the Sardinian Railways remaining. In 1880, Piedmont still had the highest number of railway stations per inhabitant. The Turin-Genoa line, which passes through the railway hub of Alessandria, is the most important. Turin, Genoa and Milan form the historic "industrial triangle", so it's natural that electrification of stations should be a priority in these regions.

Tunnelling was a necessity to enable the Italian network to communicate with other European networks. The first tunnel through the Alps, completed in 1871, was the Fréjus tunnel between France and Italy, from Modane

---

[35] The information in this paragraph is taken from Albert Schram, *Railways and the Formation of the Italian State in the Nineteenth Century*, Cambridge, Cambridge University Press, 1997.

station in Savoy to Bardonecchia in Piedmont. It is 12 km long at an average altitude of 1 300 m. Ten years later, the Gotthard tunnel was built to link Switzerland and Italy. It is 15 km long, at an altitude of 1 100 m, and links the Göschenen station to the Airolo station in the north of the canton of Ticino. It is located in Switzerland and is 53% financed by Italy, including 1/6th by the Upper-Italian Railways, according to Schram.

The electrical industry

The electrical industry expanded exponentially in Italy from the 1880s onwards. In 1882, on his return from the 1881 International Electrical Exhibition in Paris, where he had been impressed by the success of electric lighting and in particular the Edison distribution system, Giuseppe Colombo (1836-1927) founded the Committee for the Promotion of Electrical Energy Applications in Italy. After lighting Milan's cathedral square, on December 26, 1883 he and the aforementioned Carlo Monti installed electric lighting at La Scala for Almicare Ponchielli's opera *La Gioconda* and choreographer Paolo Taglioni's ballet *Flick and Flock*, using electricity generated by Edison dynamos that they had installed in the old Santa Radegonda theatre near Milan cathedral. This was the first thermal power station in Europe[36]. The distribution network was described in articles by Colombo and Rinaldo Ferrini that appeared in the French specialist magazine *La Lumière électrique* [*The Electric Light*] in 1884[37]. The Edison company for Italy, founded in January 1884 with a capital of three million lire, was the first major Italian company in the sector. Colombo, the man behind its creation, was appointed Managing Director. The company worked with numerous subcontractors to whom it delegated the public lighting of secondary towns[38]. This led to the creation of limited companies, managed by Edison or other European companies, which enabled private shareholders to raise funds for municipal lighting. The first Italian towns were lit from 1887 onwards, including Varese by the Einsteins. Electricity companies built or imported electric dynamos, whose motive power was supplied by coal- or gas-fired boilers, and installed the entire distribution network, including electricity meters. A few years later, Carlo Monti would become director of the Edison company[39].

---

[36] At the time, the driving power for electric dynamos was provided by gas or coal, and the electricity was used in the vicinity of its production site.

[37] Giuseppe Colombo, "Éclairage électrique du théâtre de la Scala à Milan", *La Lumière électrique* **11** (2), 116-117 (1884); Rinaldo Ferrini, "Éclairage électrique du théâtre de la Scala à Milan", *La Lumière électrique* **12** (5), 12-17 (1884).

[38] For a full description, see Giorgio Mori (Ed.), *Storia dell'industria elettrica in Italia*, vol. 1, *Le origini, 1882-1914*, Bari, Laterza, 1992.

[39] This company, after several mergers and restructurings, is now controlled by the French company EDF and has taken back its original name Edison.

Insurance and the fate of workers

With the development of industrialisation, which entered a second phase with the advent of electricity, new needs arose for the care and welfare of workers. Night work and child labour were more strictly regulated than in the past in the textile industry. Between 1870 and 1890, workers' mutual societies became widespread, with companies contributing to a special fund. The society for conductors on the Upper-Italian Railways was set up in Milan in 1877. The law of July 8, 1883 created the National Workplace Accident Fund (now INAIL), which managed compensation. Electricity companies, including medium-sized ones, set up their own mutual societies. Employees became members on payment of a contribution. Their role is to provide insurance cover for sickness, old age, disability, accidents at work and unemployment, and they may be backed by an insurance company to cover certain risks. Mutual societies can also play a role in saving for members and providing schooling for their children[40]. Mazzini's ideas of bridging inequalities and strengthening workers' unity were conducive to their development. In 1896, General Insurance set up the Italian Accident Insurance company in Milan, with Michele's uncle Marco Besso as director. Insurance companies diversified their activities, participating in the creation of mutual insurance companies and developing life insurance and transport insurance. On November 5, 1894, Einstein, Garrone & Cie set up its own mutual insurance company. For accidents at work, it was linked to the Patronage of Insurance and Assistance against Accidents at Work, founded in Milan in 1883 by Ugo Pisa, whose honorary president was the economist Luigi Luzzatti, whom Albert met as a minister in 1895 in Airolo.

It should be noted that the working conditions of the workers involved in building dams or digging tunnels remained very difficult. In particular, during the digging of the Gotthard tunnel between 1872 and 1881, the pace of work and the new use of dynamite led to numerous accidents and more than three hundred workers died, most of them from Piedmont and Lombardy. Three times as many died from illnesses contracted at work. A monument was dedicated to them in 1932 in Airolo. The town still bears the scars of its past. In 1877, a fire linked to the construction of the tunnel ravaged the town. It was rebuilt by moving under an avalanche corridor, which struck the town in the 1950s. Today, a huge black-and-white photograph of the avalanche is displayed on a wall in the library of the municipality of Airolo. Life for employees in the electricity companies remains just as difficult. Let's hear an account from across the Atlantic, where the development of the electricity industry began near New York with the installation of the first direct current power station in 1882 by Thomas Edison. The well-known Californian writer

---

[40] Ulisse Gobbi, *Le societa di mutuo soccorso*, Milan, Società editrice libraria, 1901.

Jack London, who was only seventeen at the time in 1893, wanted to become an electrician. His biographer Daniel Dyer reports:

> [...] as he looked around the city, he noticed an increasing use of electricity. And so, he went to the power plant owned and operated by one of the city's streetcar lines – the Oakland, San Leandro, and Haywards Electric Railway. He told the manager he wanted to learn how to be an electrician. The manager encouraged Johnny but said he must first prove himself as a shoveler of coal, the fuel that powered the electric generators for the streetcars. For thirty dollars per month, he would work ten-hour days, including Sundays, with a one-day vacation per month. He quickly learned that completing his work in ten hours was impossible. "I never finished my task before eight at night", he wrote; instead, he worked twelve to thirteen hours a day – and was not paid for his overtime [...] The strain of all the shovelling had weakened his wrists so severely that for a year afterward he had to wear supporting leather straps around them[41].

The academic world and its links to the industrial sector

Being a professor at a university or engineering school in the second half of the 19th Century was an important job. There was usually only one professor for each subject taught, supported by one or even two assistants who were renewed annually, plus a few thesis students (*allievi*) whose aim was to obtain a degree (*libero docente*), which would allow them to deliver an optional course to students. Professors were generally members of academies, the most famous of which were the Accademia dei Lincei in Rome and the Lombard Institute in Milan. Mathematicians and physicists have important links with industry. Their work involves calculations, modelling and experimental tests on industrial issues. They sit on committees set up by national or international political bodies, and are called upon to provide expert advice, including to municipalities or the State, particularly in the rail and electricity sectors. Sometimes, like Colombo, they are also directors of large companies and are active in politics, in line with their commitment to the unification of Italy. They were elected town councillors, members of Parliament, senators and sometimes ministers, and in the 1880s they moved politically towards the liberal or moderate right. They sought to protect the country's industrial revival, a symbol of growth and social progress, by defending a protectionist border policy. They therefore supported the introduction of high customs tariffs, which led to a tariff war with France in

---

[41]   Daniel Dyer, *Jack London. A Biography*, New York, Scholastic, 1997, p. 37.

1881, with farmers paying the price and pushing Northern Italy to turn to Switzerland and Germany. It was therefore hardly surprising to see German electrical companies setting up in Milan, such as the Einsteins from Munich.

Some examples of the creation of polytechnics (Turin, Milan, Rome)

With unification, the modernisation of the country went hand in hand with the creation of engineering schools. In Turin, for example, the Casati law of 1859[42] set up an engineering school modelled on the French École Polytechnique. Previously, the training of engineers had been the responsibility of the University. In Turin, students now took three years of mathematics at the Faculty of Science before specialising for two years. The Turin school is located within the walls of Valentino castle on the banks of the River Po. It was placed under the aegis of the Ministry of Public Education. A second engineering school was set up in Turin: the Industrial Museum in 1862, modelled on the École des Arts et Métiers in France, located on Corso Cavour (it was destroyed during the Second World War). It was placed under the aegis of the Ministry of Agriculture, Industry and Commerce. The two Turin schools maintained close links, with students moving from one site to the other to follow their courses. They merged in 1906 to become the Turin Polytechnic. One of the great academic figures of the time was Galileo Ferraris, the father of Italian electrical engineering, who had been a professor at the Industrial Museum since 1873. In 1885, he designed and built the first alternating current electric motor, known as a rotating field motor. He founded the Institute of Electrical Engineering in Turin, where he gave the first courses in 1888. A Turin town councillor, he was elected senator of the Kingdom of Italy and was a member of the Accademia dei Lincei. He would appear on several subsequent occasions, in connection with the Einstein and Cantoni families, Michele Besso's maternal branch.

Until 1924, there was no university in Milan. The historic university centre of Lombardy has been the University of Pavia since the 15th Century. In Milan, the Polytechnic was founded in 1863. Called the Royal Higher Technical Institute, it had around sixty engineers graduating a year, including ten electrical engineers from 1887 onwards[43]. It was founded with the help of the Society for the Encouragement of Arts and Crafts, the Lombardy Institute and the Brera Observatory, and was closer to the German polytechnic model, where teaching was more technical and less theoretical, than to the French model. Rinaldo Ferrini gave the first course in electrical engineering in 1882.

---

[42] The Casati law also made primary school compulsory in order to remedy the country's low literacy rate, of around 50% in some regions.

[43] Ferdinando Lori, *Storia del R. Politecnico di Milano*, Milan, Cordani, 1941.

The school had two sections: civil and mechanical engineering, to which architecture was added in 1865, at the instigation of the great architect Camillo Boito, from the Academy of Fine Arts in Brera. The man behind the creation of the Polytechnic, Francesco Brioschi[44], had been appointed professor of applied mechanics at the University of Pavia in 1852. He taught analysis and hydraulics at Polytechnic and directed the school, where he was succeeded as rector by Colombo in 1897. His lectures were published by one of his students, Giuseppe Jung, another of Michele's uncles. Brioschi became a senator of the Kingdom of Italy in 1865, vice-President and President of the Lombard Institute between 1868 and 1873, then president of the Accademia dei Lincei from 1884. In 1866, he took over the management of *The Polytechnic (Il Politecnico), a monthly journal of studies applied to social prosperity and culture* founded in 1839 by Carlo Cattaneo. The journal became the journal of engineers and architects. Brioschi also headed the board of the Scientific-Literary Academy in Milan, which was later chaired by Graziadio Ascoli. When the Polytechnic was founded in 1863, it was he who brought in Giuseppe Colombo, who had trained in Pavia and taught descriptive geometry at the Industrial Museum, as a professor of industrial machines, and in 1864 the economist Luigi Luzzatti. He succeeded Quintino Sella, mentioned above for his political role in connection with the railways, as president of the Accademia dei Lincei in 1884.

In Rome, the Application school for engineers, created in 1873 under the Casati law, took over from the engineering school founded in 1817 by Pope Pius VII. This delay of around ten years in comparison with Turin or Milan reflects the late attachment of Rome to Italy. When it was founded, the school was directed by Luigi Cremona, professor of graphic statics at the Milan Polytechnic, who had been appointed professor of advanced mathematics at the University of Rome. He was then replaced by Giuseppe Jung in Milan. By 1889, the Rome school saw around thirty engineers graduating each year. Guglielmo Mengarini gave the first courses in electrical engineering there in 1891 and supervised the construction of the first alternating current transmission line in Italy, between Tivoli and Rome in 1892, on which Vittorio Cantoni, another of Michele's uncles, worked. Pietro Blaserna (1836-1918), Michele's professor of experimental physics at the university in 1891, was president of the Faculty of Physical, Mathematical and Natural Sciences in 1885-1886 and in 1890-1891, the year in which he was appointed senator of the Kingdom of Italy. He was a specialist in electromagnetic induction phenomena, the basis for the operation of current generators and electric motors. He was president of the Accademia dei Lincei from 1904. His collaborator at the university was Moisè Ascoli (1857-1921), professor of

---

[44]   Andrea Silvestri (Ed.), *Francesco Brioschi (1824-1897)*, Convegno di studi matematici, 22-23 Ottobre 1997, Milan, Istituto lombardo di scienze e lettere, 1999.

technical physics and son of Graziadio. Blaserna obtained the creation of a second chair of physics at the university for Alfonso Sella, son of Quintino Sella. In 1881, he set up the Institute of Physics in via Panisperna in Rome, which would become famous for its contribution to atomic physics a few decades later (this was the institute where Enrico Fermi worked). In 1897-1898, his first assistant was Giuseppe Folgheraiter and his second assistant was Quirino Majorana (1871-1957), son of the Minister of Agriculture, Trade and Industry Salvatore Majorana. Quirino, best known for his work on wireless telegraphy, was the uncle of Ettore Majorana, the famous theoretical physicist who mysteriously disappeared at the age of 32 in 1938.

The Italian Electrotechnical Association

The Italian Electrotechnical Association (AEI) was officially founded on December 27, 1896 by Galileo Ferraris and Giuseppe Colombo. It was chaired by Galileo Ferraris, who died suddenly at the age of fifty on February 7, 1897. He was succeeded by Colombo. Guglielmo Mengarini, the man in charge of the Tivoli-Rome line, and Federico Pescetto became Vice-Presidents. It brought together industrialists, academics and professionals, and had around five hundred members, divided into six sections (Turin, Milan, Genoa, Rome, Naples and Palermo), a third of which were based in Milan. The first volume of the Association's proceedings was published in 1898 and contains around two hundred and fifty pages of articles, conference announcements, technical drawings, news from the sections, minutes of the Association's general meeting, etc. Alternating current, rotating field motors, three-phase current and electric traction remained the most frequently discussed topics within the association, following on from the pioneering work of Ferraris. The proceedings of the AEI also contain reports on international and universal exhibitions, such as the one held in Paris in 1900, which highlighted the quality of André Blondel's report on wireless telegraphy, a subject that we will link to the thesis that Michele Besso was certainly undertaking at the time. To stimulate research, the Association announced a competition with a prize of 1 500 lire for an innovative work published after 1899, with no limit on the subject, to be submitted by June 1901.

The annual congress in October 1897 was held in Milan. A visit to the Edison company was scheduled for October 25 and an exhibition of electrical equipment was set up for participants. The following day, participants visited the Pirelli and Brioschi-Finzi factories, which built the first electric trams in Milan. The AEI organised the first national congress of Italian electricians, in conjunction with the Italian Physical Society, founded in 1897 with Pietro Blaserna as its first President. It took place from September 18-23, 1899, in Como, home of Alessandro Volta. In 1900, Guido Grassi became President of the AEI, with Moisè Ascoli, Giuseppe Colombo and Stefano Pagliani as Vice-Presidents. Lorenzo Garrone, a former associate of the Einsteins, joined the Turin section as one of its three delegates. Giovanni Barberis, director of the Society for the Development of Italian Electrical Companies, which employed Michele Besso, becomes a member of the Genoa Section executive.

Hoepli and the development of Italy

Ulrico Hoepli comes from Tutwil, a village north-east of Zurich. After studying literature at the ETH in Zurich, where at the age of fourteen he was "the youngest student of Scherr's – and indeed of the Polytechnic"[45], he went to work in Trieste, where he made friends with the irredentists. In 1870, he bought the Laengner bookshop in Milan, located in the De Cristoforis gallery, near the cathedral square. He distributed works in Italian, German, French and English, and offered subscriptions to specialist periodicals.

His publishing house accompanied Italy's scientific and industrial development and in the space of a few years Hoepli acquired an international reputation. One of the first books published in 1872 was an essay by Giovanni Virginio Schiaparelli (1835-1910), Director of the Brera Observatory and professor at the Milan Polytechnic. Hoepli then embarked on a collection of technical manuals and popular science works inspired by the collection of the English publisher Macmillan, publisher of the journal *Nature*. He began with a translation of Balfour Stewart's *Physics* by Giovanni Cantoni. The most famous work in his collection was Colombo's *Civil and Industrial Engineer's Manual*, published in its first edition in 1877. The Hoepli technical collection numbered three hundred volumes in 1894. It was a response to the growing demand for training engineers, more and more of whom were leaving school, their profession becoming more widespread in Milan than that of lawyer or doctor. Hoepli was also appointed official publisher of the Brera Astronomical Observatory and of the Lombard Institute in 1872 and 1873 respectively and of the Accademia dei Lincei in Rome when Brioschi became its President. In 1878, he was also Italy's editor of the catalogue for the Paris Universal Exhibition, and published a number of fascicles on art, industry and mechanics in connection with the Exhibition, under Colombo's supervision. He also published a collection of articles by Luigi Luzzatti on pension funds and patronage. In the 1880s, he became publisher of the Royal Household. In 1886, he had electric lighting installed in his bookshop and workshops, saying: "As for the rest, the premises are very beautiful, and all the spaces are bright and large; we even have electric lighting everywhere and it works excellently; now in the evening there is a completely different light in the shop and we work much better". Hoepli drew up contracts directly with its authors, for example with Graziadio Ascoli in 1894. From 1890 onwards, Hoepli published one hundred and thirty books a year and

---

45 Enrico Decleva, *Ulrico Hoepli (1847-1935) editore e libraio*, Milan, Hoepli, 2001.

by 1900 its bookshop was selling around ten thousand books. Today it is located in the same district as it was then, in the street that now bears the name of its founder.

The young Albert frequented the Hoepli bookshop in Milan. It was located just a few steps from via Berchet, where he lived when he arrived at the end of 1894. The collection of physics lessons by Jules Violle, which is still in his personal library and on which he worked on elasticity in preparation for his admission to ETH in Zurich, bears the illustration of this publishing house on the cover. Einstein's library also contains the *Popular Lectures* of William Thomson (Lord Kelvin), published in London by Macmillan in 1889, which he may have obtained from Hoepli.

2

The company of Jakob and Hermann Einstein in the international electrotechnical context – exhibitions and reviews

The young Albert Einstein grew up in the booming world of electrical engineering. His uncle Jakob Einstein, an engineer who graduated from the Stuttgart Polytechnic in 1869, set up a water pumping and central heating company in Munich in 1876. From 1879, he was a member of the city's Association of polytechnic engineers, which brought him into contact not only with other engineers on technical matters, but also with the heads of major companies and political leaders. Hermann Einstein, his brother and Albert's father, joined him in 1880 to set up the firm J. Einstein & Cie, whose electrotechnical activities became well known from 1882. In 1885, the company specialised in this field and, with the addition of family capital, became the J. Einstein & Cie Electrotechnical Factory, which employed up to two hundred people[46]. Jakob held seven patents, relating to differential arc lamps and electrical gauges. As we have seen, from 1887 onwards the Einsteins, together with their correspondent and future partner in Pavia, Lorenzo Garrone, introduced electric lighting to a number of small Italian towns. In Munich in 1893, the company faced stiff competition for the city's street lighting contracts. At the time, there were three major companies in the sector in Germany: AEG (Allgemeine Electrizitäts-Gesellschaft), founded by Emil Rathenau in 1883, which distributed the Edison system (as did

---

[46]   The history of the Einsteins' Munich company, in the context of the development of electricity companies in Germany, is retraced by Nicolaus Hettler, whose precise details we use, in his doctoral thesis: Nicolaus Hettler, *Die Elektrotechnische Firma J. Einstein & Cie in München – 1876-1894*, Universität Stuttgart, Grin e-Book, 1996.

the Colombo company in Italy); Siemens & Halske, founded by Werner von Siemens and Georg Halske in 1847, originally for telegraphy, which patented an electric dynamo in 1867; and Schuckert & Cie, founded by Sigmund Schuckert in Nuremberg in 1873, in partnership with Alexander Wacker in 1885. After failing to win the Munich lighting contract, the Einsteins transferred their company to Italy in 1894.

We're going to look at the electrotechnical environment of the Einsteins' Munich company through the lens of the many exhibitions held around the world, which represented milestones in electrical development. The first International Electrical Exhibition was held in Paris in 1881. It was a resounding success. Railway engineer Stanislao Fadda[47] wrote in retrospect in 1883: "the Paris International Exhibition was a revelation to the whole world [...] the starting point for new studies". It certainly played a role in the very development of the Einstein company, since it was shortly before the Munich Electrical Exhibition the following year, in 1882, that the company began to develop electrotechnical activities. We will then briefly mention the Vienna Exhibition of 1883 for the measuring instruments, in particular electric meters, that were exhibited there, and we will dwell on the Turin Exhibition of 1884, marked by the presentation of the Gaulard-Gibbs transformer, which enabled alternating current to be transmitted over long distances while minimising power losses in the transmission line. It is likely that this Turin exhibition also played a role in the Einsteins specialisation in electrical engineering in 1885 and its rapid focus on the Italian market.

It should be pointed out that in the course of this chapter, the names of physicists appear who were to recur on several occasions in Albert's scientific environment: among them, the Frenchman Jules Violle, whose physics lessons he read to prepare for the ETH entrance exam in 1895; the Austrian Ernst Mach, whose critical lectures on the philosophy and history of science Albert and Michele both read; the Italian professor Galileo Ferraris, one of the fathers of international electrical engineering, to whom Albert wrote personally; the German Heinrich Friedrich Weber, who was to be his teacher at ETH, and the great scientist Hermann von Helmholtz, whose works he read. We will also say a few words about the structuring of the electrotechnical community, with its national journals in English, French, German and Italian, which enabled information to circulate by presenting the work of engineers, scientists and others. It should not be forgotten that young Albert was working in Pavia in 1895 in the office of his uncle Jakob, who made a point of acquiring these journals.

---

[47] Fadda was one of the successors to Beniamino Besso, Michele's uncle, at the head of Sardinian Railways in 1905. He had also been part of the Upper-Italian Railways since 1873.

Paris 1881, the first International Electricity Exhibition

The second half of the 19th Century saw a number of exhibitions spring up in different countries. After the first Universal Exhibition opened in London in 1851 at the Crystal Palace, a cast-iron and glass building created for the occasion and made famous by the exhibition, others were held, notably in Paris. Electricity, marginal at first, began to play an increasingly important role as a result of advances in telegraphy. The first transatlantic link was established in 1866 between Ireland and Newfoundland. Its scientific study had been entrusted to the famous physicist William Thomson (Lord Kelvin), following a first unsuccessful attempt in 1858. The copper conductors were insulated in gutta-percha sheaths using a process developed by the German industrialist Werner von Siemens. The telephone was invented ten years later. Becquerel wrote: "the discovery by Messrs Graham-Bell, in 1876, of the articulating magneto-electric telephone which telegraphically transmits speech at a distance is the most recent and certainly one of the most original of the time". Bell exhibited his system at the Universal Exhibition in Philadelphia in 1876, which celebrated the centenary of American independence and was the first in the United States. The following year, Edison's phonograph made its appearance. It was against this backdrop of the invention and development of communications that the Paris Universal Exhibition was held in 1878, for which a palace was built at Trocadéro, on the site of former quarries. Visitors were amazed by the electric lighting of the streets, department stores and the Opera house. But electricity did not yet have its own space for exhibitors, apart from telegraphy and not all its applications were represented in all their diversity.

The birth of a major exhibition devoted to electricity and its applications

The potential for the development of electricity in society prompted politicians and industrialists, looking for investors, to get together and devote a major international exhibition to it. This was the aim of the 1881 Paris International Electrical Exhibition, organised under the auspices of the Minister for Posts and Telegraphy. In the decree dated October 23 and 24, 1880 reproduced in the minutes of the exhibition[48], the Minister, addressing the President of the Council, wrote in support of the exhibition:

> Important and unexpected discoveries have recently focused the public's attention on everything to do with electricity; at the same time, industry,

---

[48] *Exposition internationale d'électricité Paris 1881, Administration-Jury-Rapports*, t. 1, Paris, Masson, 1883 (2 vols.).

seizing upon these scientific achievements, has for some years now been multiplying their applications in all fields. Today, no science seems more likely than electrical science to make rapid progress and solve problems affecting the lives of nations.

The exhibition was held at the Palais des Champs-Élysées from August 11 to November 18, 1881. It attracted a huge number of visitors: five hundred thousand paid admissions were recorded, or more than six thousand visitors a day, including invitations. In all, more than eight hundred thousand people visited the exhibition. Scientists attended reports on the work carried out at the first international congress of electricians, held during the exhibition. Industrialists, politicians and engineers mingled in the aisles and in front of the 1 700 exhibitors' stands. The themes of the exhibition were divided into six groups subdivided into classes, which are described below.

Group I: Electricity generation

    class 1. Static electricity

    class 2. Batteries and accessories

    class 3. Magneto-electric machines

Group II: Transmission by electricity

    class 4. Cables, wires and accessories; lightning conductors

Group III: Electrometry

    class 5. Electrical measuring equipment

Group IV: Application of electricity

    class 6. Telegraphy, signals

    class 7. Telephony, microphony, photophony

    class 8. Electric light

    class 9. Electric motors, power transmission

    class 10. Medical electricity

    class 11. Electrochemistry

    class 12. Precision instruments, electromagnets and magnets, compasses; electric clocks and watches

    class 13. Miscellaneous equipment

Group V: General mechanical engineering

    class 14. Generators, steam, gas and hydraulic engines, and transmissions applicable to the electrical industries

Group VI: Bibliography

> class 15. Bibliographic collections of works on electrical science and industry, plans, maps, etc.

> class 16. Retrospective collections of equipment relating to the earliest studies and applications of electricity

Each class has its own committee responsible for assessing the models on display. The competition closed on October 18, 1881. On October 21, a jury made up of one hundred and fifty members of around fifteen nationalities, half French and half foreign, awarded prizes to the exhibitors whose inventions had been singled out: fifty gold medals, two hundred silver medals and five hundred bronze medals.

Electric lighting

The omnipresent theme was electric lighting. The committee was chaired by Jean-Baptiste Dumas, Perpetual Secretary of the Paris Academy of Sciences and a politician; its Vice-Chairman was Francesco Rossetti of Padua University; and the rapporteur was Jules Violle, then a professor at Lyon University. Violle began his report as follows:

> Electric lighting has developed rapidly in recent years. It is no longer just a matter of lighting a few special places; electric light has made its way into our streets; it is a fixture in theatres, factories and railway stations; and the day is not far off when it will penetrate into our homes [...] This multiplicity of applications has given rise to numerous inventions, and the International Electrical Exhibition offers us a remarkable range of electrical appliances, from powerful lighthouses, throwing out flashes of several thousand carcels[49], to graceful incandescent lamps, the more modest equivalent of a few candles.

The use of electricity began to spread both professionally (workshops, shops, public thoroughfares, etc.) and domestically. Lamps were divided into two main categories: voltaic arc lamps and incandescent lamps.

Hippolyte Fontaine, one of the curators of the exhibition, recalls the origin of arc lamps in his account of the applications of electricity[50]:

> In 1813, fourteen years after Volta's discovery, one of England's most illustrious scientists, Sir Humphrey Davy, carried out a memorable experiment which was the seed of the new industry we are dealing with. He took

---

[49] 1 carcel corresponds to the luminous intensity produced by a Carcel-type lamp, *i.e.*, about ten standard candles.

[50] *L'Électricité et ses applications – Exposition de Paris*, Henri de Parville (Ed.), Paris, Masson, 1882, p. 129-130.

two rods of charcoal that had been previously reddened, quenched in mercury and cut with pointed ends; he placed one of these rods in communication with the positive pole of a battery and the other in communication with the negative pole of the same battery. As soon as they touched, they heated up considerably and produced a fairly bright light. On separating the two tips, the light became extraordinarily intense, and a slightly convex flame of dazzling appearance was instantly produced between the two carbon points. This flame, whose brilliance can only be compared to sunlight, was given the name electric arc [because of the shape of the flame], or, more commonly, voltaic arc, because it was obtained for the first time with the help of the Volta battery.

The problem that hindered the development of arc lamps, powered by direct current, was wear on the positive carbon: the gap between the two carbon points increased and the brightness of the lamp decreased. The Foucault regulator made it possible to maintain constant brightness by using an electromagnet in series with the lamp, which corrected the position of the carbon when it detected a variation in current intensity. But the equipment is cumbersome and reserved exclusively for laboratory use. A simpler idea is to use a solenoid through which the current flows, which magnetises a soft iron core that attracts a metal support linked to a spring attached to a coal: the Archereau device. However, in a series circuit with several lamps lined up and connected one after the other, a variation in intensity in one of the regulators disturbs the others, which is unacceptable for public lighting! *Differential lamps* solve this problem. The previous regulator is surrounded by a second solenoid with very strong fine wires, mounted in parallel with the lamp. When the length of the carbon varies, the resistance to the current changes in one of the branches, which modifies the current flowing through the second and therefore the differential action of the two solenoids on the soft iron core. Siemens presented such a system at the Berlin Industrial Exhibition in 1879. Others followed, such as the *Einstein differential lamp* patented by Jakob. It was presented by Gustave Richard in the magazine *La Lumière électrique* in 1893, where he described its mechanism in detail and set out several diagrams[51]. The "Jablochkoff candles", which had been presented at the Paris Universal Exhibition in 1878, used two parallel carbon brushes supplied by an alternating current, making their wear symmetrical.

The arc lamp, mainly reserved for industrial use, was already competing with the incandescent lamp. Violle summed up the new system as follows:

We know that a metal wire passed through by an electric current heats up, turns red and becomes incandescent; but so far, we have not succeeded in having a metal wire that does not deteriorate very quickly under the action of the current. The only substance that has lent itself to practical use is

---

[51] Gustave Richard, "Les lampes à arc", *La Lumière électrique* **49**, 313-320 (1893).

carbon but carbon is combustible in air. We therefore had to remedy this combustion, as in a regulator, or prevent it by operating in a vacuum.

He insists that "incandescent lamps in a vacuum are one of the great successes of the current exhibition". Four types of lamps were on display: Edison, Swan, Maxim and Lane-Fox. Violle points out that:

> The differences concern the substance used to make the carbon that is to form the incandescent thread; it is remarkable, however, that it is always a siliceous substance: bamboo filament, cotton with certain substances added, paper, couch grass fibre [and that only] the shape of the thread is slightly different.

The vacuum is created by pumping and exhausting air (Edison, Swan, Lane-Fox) or by substituting air with a hydrocarbon gas (Maxim). These devices made the prototypes proposed in the 1840s usable on a large scale. In the end, it was the Edison system that made the biggest impression at the exhibition, in rooms 24 and 25. Although the priority of his invention was disputed, his system won people over with "the truly marvellous combination of all the arrangements made [...] for practical lighting. Lamp holders, conduits, intensity regulators, electricity meters – everything had been planned and designed with extreme care". Not only did Edison manufacture the best quality lamps, as the comparative measurement tables published by the commission attest, but it also supplied a complete distribution system, including the generator and its power supply, cables, control and consumption measurement instruments, and so on. He was also able to advertise in the newspapers. The Edison system was subsequently developed in several countries by major companies such as AEG in Germany and Edison in Italy, headed by Colombo. Colombo himself had visited in 1882 Edison's laboratory in Menlo Park, near New York, where the first direct current generator for lighting homes had been put into operation.

Electric dynamos

Obviously, the problem of electric lighting is inseparable from that of electricity production. The first electric dynamo (generator) was produced by Pacinotti in 1864 (induction ring), then perfected by the Belgian Zenobe Gramme in 1869. In 1871, *Gramme's machine* was presented to the Paris Academy of Sciences and was marketed in 1872. Hippolyte Fontaine, whom we mentioned earlier, ran the Gramme company. The Gramme electric dynamo was powered by gas or coal, and the direct current it produced was used on site. The marketing of this dynamo was a great success, especially as Fontaine used the reversibility of the generator, which, supplied with direct current, now worked as a motor, as Pacinotti had shown. The passage of current from a generator to a motor thus enabled the "electrical transport of work". Electric dynamos of the Siemens, Brush, Meritens and Edison types were also presented in Paris. Antoine Breguet wrote: "Magneto-electric machines are very probably one of the first reasons for the existence of the present Exhibition". The electricity produced was made available to exhibitors and used for lighting by different systems, each with its own lighting ambience: the bright white light of arc lamps and that of incandescent lamps, about which Fontaine wrote that "nothing simpler, more graceful, more pleasing to the eye could be invented in the field of electric light". Initially, the idea of direct current electric dynamos was to replace batteries in order to generate significant electrical power. One of the first alternating current generators was the *Alliance machine*, designed by the company of the same name and used to power lighthouse projectors. The advantage of alternating current over direct current for the transmission of electricity over long distances became apparent three years later, with the Gaulard-Gibbs transformer presented at the Turin Exhibition in 1884, and then with the development of three-phase current and the electrical power engineering that followed.

Rail transport

There are many other sectors in which electricity is used. One of the most important was the railways, whose development was accompanied by electricity, forging close links between the two engineering communities. Electricity was widely used in this sector, primarily for signal transmission and security. *Block systems*, for example, were devices designed to prevent the simultaneous presence of two trains on the same track between two successive stations. Electric bells fixed to the front of stations warn of the departure of a train or of possible incidents, using a pre-established code. Electricity is also used to operate level-crossing alarms and to synchronise clocks at stations and throughout the country. The railways were also among the first users of telephones. Electricity was also used to light passenger carriages with incandescent lamps and to enable communication between conductors and passengers. Electric train speed controllers were introduced by the French railway company Paris Lyon Méditerranée (PLM). Electricity had been used by this company since 1877 to light the Paris station, using generators derived from those made by Gramme[52]. However, the use of purely electric traction was still in its infancy, as demonstrated by the Siemens & Halske electric locomotive at the Berlin Industrial Exhibition in 1879, which carried passengers on a ring 300 m in circumference. At the Paris Exhibition, the first overhead conductor railway line was also presented, again by Siemens & Halske. Electric railcars were not really used until the development of electric tramways at the end of the century.

---

[52] *Notice sur l'exposition des appareils électriques de la Compagnie des chemins de fer de Paris à Lyon et à la Méditerranée*, Paris, Maulde & Cock, 1881.

Medical and educational applications

The use of electricity also spread to the medical field. At the exhibition, electrotherapy uses static electricity obtained by friction, "voltaic" electricity produced by a battery and "induction" electricity from generators. Electricity is used in the curative treatment of certain nervous or muscular disorders, for example by stimulating the functioning of an atrophied muscle, or to treat rheumatism. This explains the presence of electrophysiologists such as Hugo von Ziemssen (1829-1902), director of the Munich hospital and member of the High Council of Medicine, who was one of the fifteen delegates for Germany at the congress of electricians in Paris. The physicist Hermann von Helmholtz, who had also worked on physiology, personally directed the work of the electrophysiology commission.

Educational experiments were carried out for the general public during the exhibition. For example, Ernst Mach, a professor in Prague and member of the Vienna Academy of Sciences, who was Austria's sole representative at the congress, presented apparatus for studying electrical discharges, which lead to "very curious experiments". The physics departments of Italian universities (Pavia, Turin, Genoa, Naples) made their historical collections available, as did Volta's physics department in Pavia and the Lombard Institute, which was awarded a "cooperation diploma" for the lending of its collections. A comprehensive Italian bibliography of works in the field of electricity has been compiled by the physicist Giovanni Cantoni, from the University of Pavia, and Francesco Rossetti, the exhibition's Vice-President.

The first international congress of electricians

The first international congress of electricians was held September 15, 1881, in the rooms of the Trocadéro Palace bringing together "the most illustrious electricians". The three foreign Vice-Presidents were Gilberto Govi, Hermann von Helmholtz and Sir William Thomson. The aim was to discuss measurement standards and units, so that studies carried out in different countries could be compared; the congress was therefore driven by the need for harmonisation. The first item on the agenda of the plenary sessions concerned electrical units and the "need for an agreement for the general adoption of an international system of electrical measurements". This was followed by the "choice of a system of units and the names to be given to them", with the planned setting up of an international commission attached to the International Bureau of Weights and Measures. The commission's initial work led to the introduction of units for voltage (volt), current (ampere), resistance (ohm), charge (coulomb) and capacitance (farad). The same need for harmonisation applied to telegraphy, with the introduction of international conventions and to electric lighting, with "measures to be taken to facilitate the comparison of luminous intensities".

The congress was divided into three sections for more focused discussions: section I, Physicists, chemists, physiologists: electricity from a theoretical point of view; section II, Telegraphists and railway engineers; section III, Electricians and engineers concerned with civil and military applications of electricity. Heinrich Friedrich Weber, Professor of Electrical Engineering at ETH Zurich, was the Swiss representative at the congress. Galileo Ferraris is one of the Italian representatives. In Italy he had directed the commission previously formed with Pietro Blaserna (Rome), Giuseppe Colombo (Milan), Giovanni Cantoni (Pavia), Gilberto Govi (Turin) and Luigi Palmieri (Naples) to organise Italian participation in the exhibition[53].

We have highlighted the shift by J. Einstein & Cie towards a more electro-technical field, following the Paris Exhibition of 1881. There is every reason to believe that Jakob was invited to this exhibition. He knew the Munich electrophysiologist Hugo von Ziemssen, as we shall see in Chapter 4 and also Paul Heinrich von Zech, his physics professor at the Stuttgart Polytechnic[54], who, like Ziemssen, was one of the fifteen German representatives at the congress.

Munich 1882 and the Einsteins' involvement

The first German International Electrical Exhibition took place in Munich one year after Paris. It opened its doors from September 16 to October 16, 1882 in the magnificent Crystal Palace in the Botanic Gardens, built in 1854 in the style of London's Crystal Palace. The exhibition provided an opportunity to introduce the general public to the use of electricity, to test the remote transmission of electrical power and to establish reliable measurement protocols for professionals in the sector. It was subdivided into fifteen classes: I. Historical and demonstration equipment; II. Telegraphs and signals; III. Telephones; IV. Medical equipment; V. Batteries and accumulators; VI. Electrochemistry; VII. Magneto and dynamo-electric machines; VIII. Electric light; IX. Motors; X. Miscellaneous apparatus; XI. Conductors; XII. Bibliography; XIII. Chronometers; XIV. Decorative parts; XV. Agricultural applications.

The exhibition was directed by the engineer Oskar von Miller, who had visited the Paris exhibition the year before, as Nicolaus Hettler explains in his thesis. Miller's task was to demonstrate the feasibility of remote electricity generation, with a view to the Munich municipality exploiting the

---

[53] Carlo Lacaita, "Politecnici, ingegneri e industria elettrica", in Giorgio Mori (Ed.), *Storia dell'industria elettrica in Italia*, vol. 1, *Le origini, 1882-1914, op. cit.*, p. 604.

[54] Zech was also rector of the Polytechnic School when Jakob was studying there. He gave the first course in electrical engineering there in 1882 and wrote a book harmonising German, French and English nomenclature and designations in the electrical sector.

Bavarian waterfalls by installing hydroelectric plants. For the demonstration, the mechanical power that drives the generators at the exhibition is provided by the steam engine at the city's Polytechnic and by the waterfall on the river Hirschau, which flows 5 km from the palace. On the site of the Maffei locomotive works in the Munich-Hirschau district, there was a dynamo from the Schuckert company in Nuremberg. The electric current it produced remotely drove a threshing machine at the palace.

The Munich Exhibition was marked by the construction of the first long-distance direct current electricity transmission line, between Miesbach in the south of Munich and the Crystal Palace, 57 km away. Two Gramme-type electric dynamos modified by Deprez produced the current. They were installed in the engine room of a coal mine, which provided the thermal driving power. The direct current is transmitted by two ordinary telegraph wires 4.5 mm in diameter. It was used to drive an electric pump, which propelled the water jet from an artificial fountain located in the centre of the exhibition at a height of 2.5 m. The experiment was a great success and continued throughout the exhibition. In an article published in *La Lumière électrique* in 1882, Kern described the closing of the exhibition in vivid terms, allowing us to appreciate its conviviality:

> In a sort of rag procession, like that of the students at the Polytechnic school, the whole assembly paraded, beer mugs in hand, in the presence of the President and it was not until very late at night that they parted, taking with them as good a memory of the cordiality of this party as of the Munich exhibition.

Miller asked Jakob Einstein to take direct charge of a number of the exhibition's themes. The company J. Einstein & Cie presented electrical machines such as Gramme dynamos based on the Neumayer system, to supply the current needed for electric lighting and various appliances. This dynamo, briefly presented in the German magazine *Elektrotechnische Zeitschrift* in 1883, was in operation at the exhibition alongside those of the Brush, Edison and Schuckert companies. The Einstein company also installed incandescent lamps (Swan type) and telephones (Paterson system) at the Crystal Palace and in the Munich-Hirschau district. It took part in groups III, VII and VIII of the exhibition and set up a telephone exchange with four booths lit by Swan lamps. Remote telephone transmission was a priority for the exhibition. The Einsteins installed around thirty devices, each consisting of a Paterson transfer system between two Bell telephones, at the Kils Colosseum to broadcast concerts on the exhibition site. As in Paris the previous year, the aim was to promote the telephone in a variety of ways. The Munich-Hirschau site, where the Einsteins installed their telephones and helped with the lighting, was on the site of the Maffei locomotive company. The Schwabing district, the Kils Colosseum and the Maffei locomotive company are mentioned as customers of Einstein, Garrone & Cie in the company's advertising catalogue in Pavia.

Vienna 1883 and electric meters

The Munich Exhibition was followed by the Vienna Exhibition from August 16 to November 3, 1883, which had been postponed for a year to allow its German counterpart to take place. It took place in the Rotunda, the largest circular auditorium of the time, inaugurated at the Vienna Universal Exhibition in 1873. The hall was connected to four large galleries where the exhibition was also held[55]. Fifteen thousand people visited it every day. The exhibitors, who numbered fifteen hundred, had a surface area of eight thousand square metres in the Rotunda and three hectares in total. The exhibition was divided into eight sections based on the usual themes. These include dynamo-electric and magneto-electric machines, accumulators, arc and incandescent lamps, telegraphy, telephony, electrophysiology, measuring devices, etc. The railways were represented by the Swiss locomotive factory in Winterthur. At this exhibition, the focus was on the storage of electricity by means of accumulators and on precision devices for measuring current intensity (ammeters), voltage (voltmeters), electrical resistance (ohmmeters) and energy consumption (electricity meters). As in Munich, a scientific committee tested the various devices on display. Physicist Joseph Stephan, director of the Institute of Physics and Vice-President of the Vienna Academy of Sciences, was chairman of the committee, and Galileo Ferraris was one of its nine Vice-Chairmen. Ernst Mach was one of the speakers in the exhibition amphitheatre. It should be noted that 1883 coincided with the publication of his famous work in the philosophy of science, *The science of Mechanics: a critical and historical account of its development*, which Albert Einstein read with interest in 1898.

Electricity meters proved to be a key factor in the economic development of an industrial sector that was still in its infancy. They enabled consumption to be calculated and billed to the user. One of the exhibitors was Hermann Aron (1845-1913), who was to appear in Frankfurt in 1891. In 1879, he was one of the founders of the Association of German Electrotechnicians, of which he was the first secretary[56]. Aron was a lecturer in physics (*Privatdozent*) at the University of Berlin, where he had been a former student of Gustav Kirchoff. He set up his own electric meter company following the patent he had applied for in October 1883. Michealis, in an issue of *La Lumière électrique* of November 1884, gives us a more precise idea of his device:

> Dr Aron used a pendulum clock in his new apparatus; the end of the pendulum was fitted with an iron piece and the pendulum itself oscillated

---

[55] A. Guerout, "Revue de l'Exposition de Vienne", *La Lumière électrique* **11** (5), 239-244 (1884).

[56] Shaul Katzir, "Hermann Aron's Electricity Meters: Physics and Innovation in Late Nineteenth-century Germany", *Historical Studies in the Natural Sciences* **39** (4), 444-481 (2009).

above a solenoid through which an electric current passed. This current, like gravity, influences the duration and amplitude of the pendulum's oscillations. It follows that the clock moves more slowly than an ordinary normal clock, and this delay is a function of the amount of electricity passing through the solenoid [...]. Dr Aron has combined the clock in question with an ordinary clock in such a way that the gears of both act on a counter placed between them, and the difference in the running of the two clocks can be read on the counter.

The company developed a number of subsidiaries in Europe, including one in Vienna, headed by Jakob Einstein in 1906.

Turin 1884 and the development of alternating current

The idea of an Italian General Exhibition in Turin emerged at the end of 1881, when Galileo Ferraris and Giuseppe Colombo returned from the Paris Exhibition. It was held in the magnificent gardens of Valentino Castle, home to the School of Applied Engineering on the banks of the River Po. The main entrance is located on the forecourt of the castle and visitors pass through a porch dominated by two 35 m high towers. One of these towers houses the exhibition of terrestrial and celestial physics, organised by the Astronomical and Meteorological Observatory and the Institute of Geophysics. The exhibition included several themed pavilions, representing the fine arts, music (instruments, concert halls), the *Risorgimento* Museum and an exhibition by the Italian Alpine Club, founded in Turin by the crystallographer and minister Quintino Sella. A medieval village was rebuilt on one bank of the River Po and is still one of the city's main tourist attractions. Pavilions were dedicated to the mining and chemical industries, agriculture, the navy and army, manufacturing and educational aspects (books, teaching materials). The Work gallery, the largest in terms of surface area, is divided into three sectors, one of which is devoted to electricity. The Upper-Italy Railway Company was housed in a separate pavilion alongside the electrical exhibition.

The electrical exhibition is unique in that it is the only one of an international nature within this national exhibition. It made up classes 3 and 4 of section XXII and was in fact one of the main reasons for the exhibition's existence. The Ministry of Agriculture, Industry and Commerce gave priority to "long-distance energy transport, lighting and metallurgy" and wanted to see the latest innovations in these fields presented. The themes listed in the previous exhibitions are repeated here. Energy was supplied to the generators powering the exhibition by two large steam engines, Neville & Cie from Venice and Tosi & Cie from Legnano, with 275 and 180 horsepower respectively, located in the hall of the Palace of Industry.

They powered dynamo-electrics by Pacinotti, Planté, Ganz (from Budapest), etc.[57].

The Turin Exhibition included a major innovation that filled a significant gap in previous exhibitions: the transmission of alternating current. The French physicist Alfred Potier wrote about it in retrospect in 1889:

> Ever since Gaulard's remarkable experiments at the Turin exhibition drew public attention to the ease with which alternating currents can be transformed, people have been thinking of taking advantage of this to establish long-distance power transmission by sending a low-intensity current along the line, transformed in the vicinity of the receiving machine so that it no longer produces unacceptable voltages. The use of these currents is subject to the discovery of an alternating current motor that is as easy to start as a direct current motor; the complete solution to this problem does not yet seem to have been found, but approximate solutions have been given that it would have been interesting to see in action [birth of the three-phase current].[58]

The transformer, then known as the Gaulard-Gibbs "secondary generator", was about to revolutionise the field of electrical engineering. It was developed by the Frenchman Lucien Gaulard (1850-1888), who joined forces with the Englishman John Dixon Gibbs (1834-1912) in 1883 to bring it to market. Until then, electric lighting could not compete with gas lighting within a radius of more than 500 m, because of power losses in the line. The transformer was presented by the London-based National company for the distribution of electricity by secondary generators. Its aim is to supply incandescent lamps, arc lamps, motors, etc. remotely and economically. For its demonstration, it used a 100 Hz alternating current produced by a Siemens dynamo located on the exhibition site and powered by the Tosi steam engine. It delivered a voltage of 3 kV and a current of 11 A. Two conductors carried the electricity *via* the main station at Porta Nuova in Turin to Lanzo, 17 km away. They were attached to the telegraph poles of the Upper-Italian Railways. The round trip from the exhibition site is around 80 km. Transformers placed at various points along the line allowed electricity to be used to light lamps that operate at low voltage. Edison streetlamps and Swan and Bernstein lamps were supplied to the exhibition site, Lanzo station and intermediate stations. This was the first time that electricity has been transmitted over such a long distance with an efficiency of around 50%, between the power consumed by the lamps and the apparent power produced. As today's physics students know, in order to transport current without excessive losses due to the Joule effect (overheating), it has to be transported at high voltage (and therefore

---

[57] *L'ingegneria, le arti e le industrie alla esposizione generale italiana in Torino 1884*, Rivista tecnica compilata, edited by engineer Giovanni Sacheri, Turin, 1890.

[58] *Exposition universelle internationale de 1889 à Paris, Rapports du jury international, Alfred Potier, 62 – Électricité*, Paris, Imprimerie nationale, 1891, p. 12-13.

low current). The Gaulard-Gibbs transformer is used to step up the voltage at the output of the generator, while other transformers are used to step down the voltage to provide the user with a voltage suitable for his/her appliances. Their principle is based on the phenomenon of magnetic induction[59]. An article by Giuseppe Colombo in *La Lumière électrique* in 1884 describes the system used. The transformer was presented and studied from a theoretical and practical point of view by Galileo Ferraris, who chaired the electricity session at the exhibition and reported his findings to the Turin Academy of Sciences on January 18, 1885. In order to carry out his studies, Ferraris designed and built new measuring instruments, particularly for power, to take account of the phase shift between current and voltage: this involved determining the "cos Φ" parameter, which enabled the efficiency of the device to be calculated and the power consumed to be quantified. Ferraris thus established that the power efficiency of alternating current is greater than that incorrectly obtained using direct current calculation techniques. At the exhibition, Gaulard and Gibbs were awarded 10 000 lire by the city of Turin for their major contribution to electrical development. The city council opted for alternating current lighting, splitting the contract between Gibbs and Edison. Ferraris proposed using the Gaulard-Gibbs transformer in hydroelectric power stations. The first transformers were installed at the Tivoli power station, near Rome, under the direction of Vittorio Cantoni, the maternal uncle of Michele Besso, an engineer from ETH Zurich and Milan Polytechnic, whom we will meet again in the next chapter.

The exhibitions are ideal venues for exchanges between new generations of engineering students. In Turin, for example, Italian engineers and professors welcomed student engineers from ETH Zurich, accompanied by their professor of graphic statics. The illustrated chronicles in the exhibition recount how a delegation of one hundred and fifty ETH students from the Gotthard were received by the Turin engineering students, with banners, hat-throwing and vivats chanted as they arrived on the platform at Porta Nuova station. During their four-day stay, May 21-24, they visited Turin, the Application school on the Valentino site and, of course, the International Electrical Exhibition, which had opened on May 20 and where they met again to end their stay. They attended a lecture by Professor Rinaldo Ferrini, from Milan Polytechnic, on the remote electrical transmission of information from measuring instruments.

The Einsteins probably visited the Turin Exhibition, before specialising in electrical engineering and investing in the development of their new Munich company in 1885. It is worth noting that they were in contact with political and industrial leaders in Northern Italy before 1887. Galileo Ferraris's personal archive contains two letters from Jakob Einstein on company letterhead

---

[59] The variation over time of the flow of the magnetic field through an electric circuit causes an induced current to appear in the circuit.

paper. They concern a disagreement with the engineers Battisti, Castiglioni and Gorla over the lighting of the town of Varese, which had been decided in July 1887. The first, dated January 26, 1888, sets out the reasons for the dispute. The Einsteins thanked Ferraris for his reply to their first letter and his agreement to intervene[60]. Clearly, they would not have written to him if they had not had the opportunity to meet him beforehand, at the Turin Exhibition for example. It should be remembered that the Einsteins were working in Italy with the Turin engineer Lorenzo Garrone, a young graduate of the Turin School of Engineering (class of 1881).

Frankfurt 1891, three-phase current and the Einsteins' participation

The Frankfurt Exhibition was set up in response to a request from the city's elected representatives in 1887, invited to give their opinion on the various systems of electric lighting and current transmission, whether direct or alternating, proposed to the municipality. Ferraris was one of the three personalities initially approached to provide an expert opinion. The Frankfurt Exhibition is famous for the inauguration, in August 1891, of Europe's first three-phase alternating current power line (178 km). It linked Frankfurt and Lauffen am Neckar, where hydroelectricity was produced by the river. A commission met to compare the electrical systems and equipment presented and assess the line's performance. Chaired by Hermann von Helmholtz, it was made up of nine groups. The committee's report on the 1891 Exhibition can be found in the Reports on the Work of the Examining Committee[61]. Weber, Professor of Physics at ETH Zurich, was in charge of Group IX on the transmission of electric current.

At the same time as the exhibition was taking place, a first political congress was held from August 27-29 to discuss the situation of municipal electric lighting and long-distance power transmission in Germany. It was attended by six hundred delegates, a third of whom were foreigners. A scientific congress followed September 7-12, bringing together electrical engineers. Hermann von Helmholtz had assigned the chairmanship to Galileo Ferraris, acclaimed by the audience as the father of theoretical electrical engineering. The three-phase current used on the Lauffen-Frankfurt line in

---

[60] These letters, written on the letterhead of the Elektrotechnische Fabrik J. Einstein & Cie, are listed in *L'archivio di Galileo Ferraris*, vol. 1, *Corrispondenza-Inventario*, edited by Raffaella Gobbo and Andrea Silvestri, Vercelli, Gallo, 1997, which also contains a biography of Ferraris.

[61] *Offizieller Bericht ueber die internationale elektrotechnische Ausstellung in Frankfurt am Main 1891*, Bd. II, *Vorstand der internationalen elektrotechnischen Ausstellung in Frankfurt am Main*, Frankfurt, Sauerländer Verlag, 1894.

fact put Ferraris' proposed system for power electrical engineering into practice on a large scale. In his inaugural lecture for the year 1899-1900 at the Turin School of Application, Guido Grassi, President of the Italian Electrical Association, recalled that at this congress the engineer Karoly Zipernowski, from the Ganz company in Budapest, had proposed a high-speed electric traction system (250 km/h) using four 200-horsepower direct-current motors, a project which in his view deserved to be rethought with a Ferraris rotating-field motor[62].

Like the Turin Exhibition, the Frankfurt Exhibition was a forum for exchanges between engineering students and Ferraris was accompanied there by a large delegation of Italian Polytechnic engineers. It should be pointed out that the many patents (which did not include Ferraris) on three-phase motors were taken out shortly afterwards in the United States by the Serbian-born engineer Nikola Tesla (1856-1943), who had gone to work for Edison before founding his own company and joining Westinghouse.

After Munich, the Einsteins took part in the Frankfurt Exhibition. It was probably on this occasion that the young Albert was introduced to Galileo Ferraris. We will see in Chapter 4 that in 1895 Albert asked him for a letter of recommendation to attend Weber's classes at the ETH in Zurich. This exhibition was an important moment in the life of the twelve-year-old Albert, who became fully aware of the electrotechnical environment in which his family was evolving and could not help but marvel at it. The Einsteins presented the activities of their company, which manufactured electric dynamos and arc lamps, marketing incandescent lamps and other electrical appliances. Their company also transported electric current to power DC machines and for public lighting. They provided lighting for a number of public places, including the Milani café, the beer hall, the labyrinth and the shooting range[63]. Their lamps were tested by Group II of the exhibition committee in a comparative study with five other lamps. The Einsteins also presented an electric meter system designed by their engineer, Sebastian Kornprobst, which was described in December 1891 in the *Elektrotechnische Zeitschrift*. Their company took part in the symposium on electric lighting alongside twenty others, including Hermann Aron's electric meter company, which presented the model used for the three-phase current on the Lauffen-Frankfurt line. It was certainly on this occasion that Jakob Einstein met the man who was to become his future employer in 1906, unless he had already met him in Vienna in 1883. Jakob gave a presentation at the congress in which he "described the electrical distribution system that the company had set up in the Munich quarter of Schwabing, as well as in the small, northern Italian

---

[62] Vittorio Marchis (Ed.), *Letture Politecniche I (1889-1906)*, Turin, Centro Studi Piemontesi, 2008.

[63] *Eine neue Zeit! Die internationale elektrotechnische Ausstellung 1891*, Frankfurt, Frankfurt am Main Historisches Museum, 1991.

towns of Varese and Susa[64]". An article in *La Lumière électrique*, based on an article in the *Centralblatt für Elektrotechnik*, gives a detailed description of the lighting system installed in Schwabing by "Messrs J. Einstein & Cie[65]". After dwelling on the originality of the installation for supplying gas to the turbine of the generator – one of Jakob's first specialities – the article tells us that:

> The lighting consists of 170 16-candle incandescent lamps, 30 32-candle lamps and 10 8.5-amp arc lamps. [...] suspended in the middle of the streets on cables stretched across them. For service, they are lowered to the ground using a running trolley system with counterweights. The incandescent lamps are carried on arms and encased in a glass bell.

Following the loss of the Munich public lighting contract in 1893, the Einsteins turned their attention to Italy, where they settled in 1894 at the initiative of Jakob and their correspondent Lorenzo Garrone[66] to set up the firm Einstein, Garrone & Cie.

Electrotechnical magazines

In the wake of the first international electricity congresses, the 1880s saw the emergence of several national electrotechnical journals, specific to the new community that was developing. There were many links between these different journals, particularly through articles and news briefs (production figures from major companies, awarding of contracts, news from correspondents, etc.), translations of academic articles deemed important, notes on scientific works, announcements and reviews of exhibitions held in the sector around the world, which enabled information to circulate easily from one country to another.

The oldest magazine is the English weekly *The Electrician*, subtitled *A Weekly Illustrated Magazine for Electrical Engineering, Industry and Science*. It was founded in 1861 and published in London. It ran to around one thousand four hundred pages a year. Among the scientific articles it published were the proceedings of the Royal Society, whose debates it also reproduced, such as the discussions in 1892 between FitzGerald and Lodge on ether drag, which will be discussed in Chapter 5.

In France, Cornelius Herz's weekly *La Lumière électrique* was founded in 1879. The magazine advertised itself as a *Universal electricity journal*.

---

[64] Lewis Pyenson, *The Young Albert Einstein. The Advent of Relativity*, Bristol, Adam Hilger, 1985, p. 45.

[65] "L'installation d'éclairage électrique de la ville de Schwabing, près Munich, en Bavière", *La Lumière électrique* **33**, 132-134 (1889).

[66] "Albert Einstein - A Biographical Sketch" (Translated Excerpts) by Maja Winteler-Einstein in *CPAE (English translation supplement)*, vol. 1, p. xvii.

From 1894 onwards, it became *L'Éclairage électrique* [*Electric lighting*], edited by Jules Blondin, and was clearly the reference in the field for the international community, with four six-hundred-page volumes published each year.[67] In alphabetical order of the table of contents, the main subjects covered were: accumulators, alternators, electricity meters, electricity distribution, direct-current dynamos, city lighting, electric tramways, the use of waterfalls for electricity production, etc. The special feature of the French magazine was that it made available to the engineering public the fundamental work underlying the electromagnetic theories of James Clerk Maxwell, Heinrich Hertz, Hermann von Helmholtz and Hendrik Antoon Lorentz, for example, written by Editor-in-Chief Jules Blondin, Alfred Liénard, professor at the Saint-Étienne School of Mines and even the famous mathematician and theoretical physicist Henri Poincaré. In 1893, for example, following on from Gustave Richard's article on differential arc lamps and Jakob Einstein's system, there was a paper on the principle of least action in Helmholtz's theory, a theory that would be of interest to Albert in 1897. It should be noted that the electricians' magazines also echoed the controversies that arose over the priority of patents. For example, *La Lumière électrique* published the letters that Lucien Gaulard exchanged with other engineers about the paternity of the transformer[68]. The controversy led to his defeat, his ruin and his early end in a psychiatric asylum.

The German journal of the Association of Electrotechnicians, founded in 1879, was the *Elektrotechnische Zeitschrift*, created in 1880 at the instigation of Werner von Siemens, a member of the Royal Academy of Sciences in Berlin. It was edited by Rühlmann and Petsch and printed by Julius Springer in Berlin and Oldenbourg in Munich. The first issue contained 440 pages for the year. This fortnightly magazine featured a large number of high-quality diagrams and photographs of electrotechnical devices. The calculations presented were primarily practical for engineers. In 1890, the magazine changed format, adopting a larger page layout certainly in line with the care taken in presenting the technical drawings and reproductions. Under Uppenhorn's editorship, it became a little closer to the French magazine, although it did not venture so far into the realm of theoretical presentations. This difference in approach reflects the difference in the more theoretical training of French engineers at the École Polytechnique and the more practical training at German Polytechnics.

The magazine *Elettricità*, published by Lamperti in Milan, was created in 1886. It took the form of a sixteen-page illustrated weekly magazine. It contained technical descriptions, mentioned theoretical work in conjunction

---

[67] See for further details Nicolas Nio, *Electromagnetic Theories of the Ether: their French diffusion, particularly in technical higher education and journals dedicated to electricity at the end of the 19th Century* (in French), Thesis, Paris Observatory (2020), pp. 261-377.

[68] Lucien Gaulard, *La Lumière électrique* **11**, 136 (1884).

with academies, reports on patents, publications and exhibitions, and included a few short articles but relatively few calculations. It also contains advertisements from manufacturers and electrical contractors. The oldest Italian journal is the Florentine *Elettricista*, founded in 1877 and edited by Lamberto Cappanera, who translated articles from the English journal *The Electrician*[69]. The proceedings of the AEI published in 1898, which we mentioned in the previous chapter, evolved in 1914 into a genuine periodical devoted to electrotechnology, *L'Elettrotecnica*, in which the best-known Italian scientists, engineers and industrialists published[70].

As we have said, these magazines were to be found in Jakob Einstein's office, where he kept abreast of the latest advances in the field and market trends: the *Elektrotechnische Zeitschrift* and *La Lumière électrique*, because they were the most important on the continent and also detailed his own inventions; *Elettricità*, because it was the Milanese magazine that was useful for more local information; *The Electrician*, because it was the pioneering magazine in the sector and provided an insight into the Anglo-Saxon world. Remember that English, German and French are the three languages used in the training of European engineers.

The young Albert, aged sixteen in 1895, certainly read extracts from these magazines out of curiosity and built up his knowledge of electrotechnology while working in his uncle's office and preparing for the ETH Zurich admission exam. In particular, the effect of magnetic induction, which is the basis of the functioning of the electric dynamos developed and marketed by his uncle. It should be remembered that it was by going back over the two distinct interpretations of this effect – which results in a current, Lorentz's induction or Neumann's induction, depending on whether an electric circuit is moving in a constant magnetic field or whether the circuit is fixed in a variable magnetic field – that Albert introduced in his article of June 1905 the need for a single explanation by virtue of the postulate of relativity; an introduction that was certainly also a form of homage to the family's electrotechnical environment.

---

[69] Roberto Maiocchi, "La ricerca in campo elettrotecnico", in Giorgio Mori (Ed.), *Storia dell'industria elettrica in Italia*, vol. 1, *Le origini, 1882-1914*, *op. cit.*, pp. 155-199, here p. 157.
[70] It still exists today under the association's current name, AEIT.

3
Michele Besso and his family's role in Italian industrialisation

Michele Besso's role with Albert Einstein was both very discreet and very present. He was his closest collaborator and best friend throughout his life. After becoming friends in Switzerland in 1895-1896, Albert and Michele met again between 1899 and 1901 in Milan during the university holidays; their friendship was consolidated by almost daily scientific discussions, before they met again in Bern at the beginning of 1904. In the introduction, we mentioned Michele's participation in Albert's work on quanta in 1904 and special relativity in 1905, followed by general relativity in 1913-1914 in Zurich. However, their scientific interaction goes back much further and has evolved over time. We will raise the question of its origin, still undetermined, which could date back to Albert's first attempt to enter the ETH in Zurich at the age of sixteen, in October 1895.

One of Michele's essential traits was his attachment to a humanist culture, marked by a passion for all forms of knowledge, whether scientific, philosophical, economic, social or literary. To gain a better understanding of his personality, we need to go beyond the information provided by Pierre Speziali in his introduction to the Einstein-Besso correspondence, from which we will nevertheless borrow some valuable information. We will be looking at the influence that not only his father Giuseppe had on him, but also his uncles Marco and Beniamino, who held important management positions in the insurance and railway industries, and his uncle Davide, a mathematician. We should also look at his maternal line, which has been little studied until now, with his uncle Vittorio Cantoni, a civil engineer who, like him, graduated from ETH Zurich, and Giuseppe Jung, who played an important role in Albert and Michele's scientific environment in Milan. This discussion will also give us a better understanding of the context in which the two friends met. In his life-time Michele devoted himself to finding a balance, based on his love for family and for knowledge. Assisting and participating in the development of science, alongside Albert, was one of his greatest pleasures.

Michele Besso (1873-1955), Albert's long-standing friend

A brilliant young man

Michele Angelo Besso was born in Riesbach near Zurich on May 25, 1873, according to Pierre Speziali. He was therefore six years older than Albert. He probably learnt to read from his uncle Beniamino's book *Great Inventions and Discoveries*[71]. Speziali reports that Michele was "a precocious child, curious about everything". He quotes him in 1953, when he says that he "learned to read in his fifth year, from a book that revealed to him, in physics, technology and astronomy, almost everything that could have been known without mathematical formulae ninety years ago". The first edition of Beniamino's work dates from 1864, *i.e.*, eighty-nine years before this quotation. It's easy to imagine that the thoughtful uncle gave this richly illustrated book to his nephew as a token of affection, but also to give his inquisitive mind something to marvel at and reflect on. However, this episode coincided with a traumatic event for Michele. His mother, Erminia, became seriously ill and was no longer able to look after him. It would not be surprising if the discovery of this book at this crucial moment in the development of Michele's personality was in some way the starting point for the unquenchable thirst for knowledge that would drive him for the rest of his life.

At the age of sixteen, Michele was expelled from his grammar school in Trieste following a petition he had circulated with a classmate against his maths teacher, whom he considered incompetent. He then enrolled in the first year of the Faculty of Science at the University of Rome and moved in with his uncle Beniamino. His student file in the University archives states that he took the mathematics course as an "apprentice engineer", attested by the "Matura [Swiss baccalaureate] issued on July 9, 1890 by the Lyceum Comunale Superiore of Trieste, and recognised as equivalent by the faculty". He obtained excellent marks in scientific subjects: 28/30 in analytic and projective geometry with Alfonso del Re; 30/30 in algebra with Luigi Biolchini; 29/30 in experimental physics with Pietro Blaserna, whom we have already met as Rector of the Faculty of Science; these contrasted with a more modest 20/30 in geometric drawing. To give a relative indication of the quality of these results, it should be pointed out that the physicist Quirino Majorana, whom we have already mentioned and who was Blaserna's assistant for several years before becoming professor of experimental physics at the Politecnico in Turin and succeeding Augusto Righi at Bologna in 1921, had obtained marks rather close to 20/30 in the first year of his degree in Rome in 1887.

---

[71] Beniamino Besso, *Le grandi invenzioni e scoperte antiche e moderne nelle scienze, nell'industria e nelle arti*, Milano, Fratelli Treves, 1864.

However, Michele did not continue his studies in Rome, where he could have specialised, like Quirino, at the Engineering application school, founded in 1873 and directed by Luigi Cremona, a colleague of his uncle Giuseppe Jung; instead, he turned to the ETH in Zurich, which he joined immediately, in 1891. He graduated on March 14, 1895 in mechanical engineering, just as Albert was preparing for the entrance exam. The minutes of the deliberations state that the students "Besso and Vonnex obtained insufficient marks in the graphic examination part of their diploma. However, in view of their excellent marks in the final oral exam and the intermediate exam, both are recommended for the degree[72]". Michele obtained the maximum mark of 6 in mathematics with Adolf Hurwitz, as well as in mechanical engineering with Aurel Stodola, in experimental physics and electrical engineering with Heinrich Friedrich Weber and a 5.5 in physics with Weber. He kept in touch with Stodola, who introduced him to Ernst Mach's critical lectures on physics, to which he introduced Albert in 1898, as we shall see in Chapter 6. Michele also took optional courses, including Albin Herzog's on elasticity and Weber's on three-phase currents, as well as courses in French political economy and English literature.

Constant links with Albert

Speziali writes that "shortly after presenting his engineering diploma work, Besso obtained a position at the Rieter electrical machinery factory in Winterthur". In 1898 he married Anna Winteler, the eldest of the seven children of Jost Winteler and his wife Pauline Henriette Eckardt, who had taken in Albert Einstein during his school year in Aarau in 1895-1896. Jost Winteler (1846-1929) had completed his dissertation in linguistics with Eduard Sievers in 1875, a work regarded almost a century later as pioneering in its field by the famous linguist Noam Chomsky. He taught history and Greek at the cantonal school in Aarau from 1884 to 1914 and tried his hand at poetry: he wrote *Tycho Pantander*, published in 1890 by the Huber bookshop in Frauenfeld, a book read by Michele and Albert. The Einstein and Besso families were linked *via* the Winteler family, since Maja Einstein, Albert's younger sister, married their son Paul. Family and a thirst for knowledge were the driving forces in Michele's life, so it's hardly surprising that he thanked Albert on January 17, 1928, with these words[73]:

> I owe you my wife and thus my son and grandson; I owe you my position, and along with it the peace of the worldly cloister, and my financial security in hard times. I owe you the scientific overview that without such a friendship can be achieved only at the expense of one's whole person, if at all and you know better than anyone what a powerful, supra-individual feeling of life is connected with that.

---

[72] Communication from the ETH Zurich archives.
[73] Letter from Michele Besso in *CPAE (English translation supplement)*, vol. 16, Doc. 132, p. 141.

He recalls the date of his meeting with Anna Winteler when, in a letter in French dated May 25, 1945, he informed Albert of his wife's death: "You know that on September 22 I lost the companion of half a century whom I met through you in June 1897". Michele and Anna had an only son, Vero, born in 1899.

After Winterthur, Michele Besso moved to Milan in 1899, according to Speziali:

> Besso went to work in Milan for a company that transmitted electrical energy over long distances. From 1900 to 1901, Michele was a technical adviser to the Society for the Development of Italian Electrical Companies. During these two years, he played an active role in drawing up safety measures for electrical installations, as laid down by the Italian insurance companies.

When his father died in October 1901, he moved to Trieste, from where his paternal family hailed originally. From 1904, he joined Albert in Berne at the Patent and Intellectual Property Office headed by Friedrich Haller, where he shared his daily scientific questions on quantum physics and relativity. In 1909, he attended courses and lectures in a wide variety of fields, at a time when Albert had become a professor at the University of Zurich. We then find Michele in Gorizia:

> Family circumstances and interesting business prospects led Besso to move with his wife and son to Gorizia, near Trieste, where he had just been offered a position as technical adviser in a textile factory. He soon found himself a member of the management of this company and, at the same time, a member of the board of directors of an insurance company.

He then returned to Switzerland, of which he was a national, when Italy entered the war in 1915 and wanted to retake Trieste and Trento from the Austrians in order to complete unification. After briefly teaching physics at the Technical College in Winterthur, he lectured on industrial law and the protection of inventions at the ETH in Zurich from 1916 until his retirement in 1938 and returned to the Patent Office in Berne in 1919. Speziali describes Michele as a passionate and tireless reader, who attended numerous conferences and worked in the library at the University of Geneva, even towards the end of his life. Their correspondence continued after Albert had emigrated to the United States and touched on the problems of their time and the political, social and economic issues at stake. Michele asked his friend questions about his latest work, for example in July 1954 about the effect of a rotating electric charge on the structure of the equations of the unit field theory on which Albert was working.

An endearing personality

As we have already said, Michele Besso is an eclectic man, interested throughout his life in the sciences, mathematics, philosophy, law, economics, medicine, psychology, etc. His concern for social issues is inherited from a family sensitivity and culture. His father Giuseppe and his uncle Marco had a humanist political vision and as insurance managers they worked to improve working conditions and care for workers and their children, alongside financial investment in the development of the industrial sector, which was also synonymous with social progress at the time. Michele's main aim was to improve his knowledge through constant hard work. So, from a professional point of view, he seems less passionate about the practical aspects of the engineering profession. Albert was amused by this, while making no secret of his deep esteem for his friend's intelligence. In a letter dated March 27, 1901, he tells Mileva that Giovanni Barberis, "Michele's director, with whom we are rather well acquainted, was at our house for music making". He then wrote: "He said how totally unusable and almost mentally incompetent [not responsible for his actions in a legal sense] Michele is, despite his extraordinarily extensive knowledge[74]." He went on to say that "Once again, Michele had nothing to do", and went on to tell an anecdote in which he was seen to miss one train after another on his way to the power station he was to inspect. If this passage is anything to go by, Michele's position as adviser gave him precious free time, which he needs for his scientific reading. For his part, Albert, who had grown up in an environment of entrepreneurs struggling to survive on a daily basis and who had his sights set on a scientific career that would be impossible to pursue without support, sometimes found it hard to accept his friend's self-lessness. On March 27, 1901, for example, when he told Mileva that he was going to "try to get an assistant's position in Italy", he immediately pointed out that he has "quite good connections here", including that of Giuseppe Jung, "Michele's uncle [...] a professor of mathematics", whom he described in the following letter as "one of the most influential professors of Italy". But having few illusions about Michele's eagerness to ask his uncle for help, he immediately adds: "I'll grab him by the scruff of his neck and drag him along to his uncle & once I am there, I'll do my own talking". Albert waited impatiently for Michele, who had left for Trieste, to return. Two days after seeing him again, he writes to Mileva on April 4, 1901, clearly still very annoyed: "Michele arrived with wife and child from Trieste the day before yesterday. He is an awful weakling without a spark of healthy humaneness, who cannot rouse himself to any action in life or scientific creation". But he immediately corrects himself, adding: "But an extraordinarily fine mind, whose working, though disorderly, I watch with great delight. Yesterday evening I talked shop with him with great interest for almost 4 hours". Albert goes on to give us a remarkable insight into the wealth of scientific subjects discussed with Michele: "We talked about the fundamental separation of luminiferous ether and matter, the definition of absolute rest, molecular forces, surface phenomena, dissociation". We can imagine that in their discussions Michele,

---

[74] Letter to Mileva Marić in *CPAE (English translation supplement)*, vol. 1, Doc. 94, pp. 160-161.

who was more encyclopaedic, must have been critical of the young Albert's ideas. Albert, on the other hand, was keen to pursue the path that attracted him, even if it meant realising afterwards that he had gone down the wrong road (the "bad leads" referred to by Michele). Michele confirms this in a letter to Albert in 1932, in which he mentions among the links that unite them "my youth and yours, the time when your genius brought you discoveries in batches, from which you had, with effort and tenacity, to extract the one that was valid and the pure joy that I felt, and my countless objections of all kinds [...]". Their difference in character must have annoyed Albert at times, who goes on to say in his letter to Mileva that Michele "is very interested in our investigations, even though he often misses the overall picture because of petty considerations. This is inherent in the petty disposition of his being, which constantly torments him with all kinds of nervous notions". In 1948, he illustrated this difference by depicting himself as a "mole" and Michele as a "butterfly", concluding: "A butterfly is not a mole, but no butterfly should regret it". However, as we said in the introduction, Michele was the best sounding board in Europe for him.

Albert and Michele meet

What are the circumstances that lead to the two friends meeting? When did it take place? The answers to these questions will help us to better understand Michele's role and attitude towards Albert. For the moment, we can only speculate, given the partial, contradictory and subjective information surrounding this meeting. In the letter Albert wrote to Vero Besso after Michele's death in 1955, he reminded him of his father: "Our friendship began when I was a student in Zurich". In Albert's eyes, their friendship *a priori* dated from October 1896. But the date of their first meeting was probably earlier. It was in October 1895 that Albert arrived in Zurich to apply for admission to the ETH. In his *Autobiographical sketch*, he writes: "In 1895, at the age of sixteen, I arrived in Zurich after spending a year with my parents in Milan without a school or teacher. My goal was to take the entrance exam to the Polytechnic, but it was not clear to me how to go about it[75]". Michele referred to the year 1895 to date their first scientific discussion, in a letter to Albert dated October 10, 1945: "Isn't this again our starting point of more than fifty years ago, between Newton and Huygens[76]?", a precise allusion, given Michele's "fussy" nature. Perhaps he even intended it to commemorate

---

[75] Albert Einstein, "Autobiographische Skizze", in Carl Seelig, *Helle Zeit – Dunkle Zeit*, Zurich, Europa Verlag, 1956, pp. 9-17. Translation from the original German.

[76] Translation based on the letter reproduced by Speziali in the German edition of the Einstein-Besso correspondence of 1972. In the French translation of 1979, Speziali erroneously wrote "our starting point of fifty years ago", omitting "more than" which is essential for dating the Einstein-Besso meeting.

the fiftieth anniversary, between October 1895 and October 1945, of what he considered to be their first meeting? This remark also clearly shows that their relationship began on a scientific basis, suggesting that Michele, who was six years older than Albert, played a very early role in the young Albert's scientific training.

As to where they met, Albert asked Michele in 1952 to tell Seelig "after we had met at Madame Caprotti's". Selina Caprotti, whose maiden name was Hüber[77], was married in Zurich in 1871 to Carlo Caprotti (1845-1926), director of the important Caprotti spinning mill in Ponte d'Albiate, north of Milan. Michele's presence in her home seems to involve the maternal branch of her family and the textile industry. We shall see that Angelo Cantoni, Michele's grandfather, was a major shareholder in the Filatura Cantoni owned by Baron Eugenio Cantoni in Legnano, who himself had a professional relationship with the Caprotti family[78]. Carlo Caprotti and Selina Hüber were also of the same generation as Michele's parents, Giuseppe Besso and Erminia Cantoni, and the couple married a year apart in Zurich. If Michele's presence in the Caprotti household can be interpreted quite naturally, how did Albert enter their circle? Was it through the Marangoni family in Pavia, who raised silkworms for the textile industry? Or was it because the Caprottis were friends of Albert's acquaintances in Switzerland, such as Gustav Maier, a former banker and head of a weaving company in Zurich, who took charge of Albert when he arrived in 1895 and then placed him with his friend Jost Winteler in Aarau[79]? Or Rudolph Einstein, his father Hermann's cousin, who had made a fortune in textiles in Hechingen and helped finance the Einsteins' business? Albert's recollections, passed on by Michele to Carl Seelig, become confused under the latter's pen[80]. Speziali's precision of date and place appears contradictory. In the foreword to the Einstein-Besso correspondence, he places the meeting "in Zurich in 1897", when "Einstein was 18 and Besso 24". But in the introduction that follows, he explains: "It was 1896, early autumn. Besso, who often travelled to Zurich, frequented a family of music lovers, the Hüni, where he played the violin. It was during one of these evenings – the exact date has not been established – that he *first met Albert Einstein* [italics in original]. He was now 23, whereas

---

[77] The surname Hüber appears several times in the Einstein-Besso correspondence, in particular Eugen Hüber, "creator of the Swiss Civil Code".

[78] One of his collaborators, Giuseppe Frua (1855-1937), moved from one company to another.

[79] Correspondence between Gustav Maier and Jost Winteler in October and December 1895 in: *CPAE (English translation supplement)*, vol. 1, Doc. 9, p. 8 (Letter from Gustav Maier to Jost Winteler) and Doc. 13, p. 10 (Letter from Jost Winteler to Gustav Maier).

[80] He wrote: "Michele Angelo Besso, whom Einstein had met through Anna Winteler – Besso had married the eldest daughter of this academic couple when he was living in Aarau – in Zurich in a lady's sitting room".

Einstein, who was born on March 14, 1879, was only 17". There is also no further mention of Caprotti.

It's worth noting that there are many links between the Einstein and Besso-Cantoni families, apart from the textile industry mentioned above. Many others are intertwined: Jakob Einstein and Vittorio Cantoni, both Polytechnic engineers, had links with the father of electrical power engineering, Galileo Ferraris; the workers' mutual societies in Lombardy companies, including that of the Einsteins, were backed by insurance companies that managed the risk of accidents at work, such as the one set up by Marco Besso in Milan; the railways, in which Beniamino Besso, after helping to excavate the Fréjus tunnel, took a leading role, were linked to the electrical industry, where the Einsteins were involved in lighting several stations. The Einsteins also had political connections, such as the minister Luigi Luzzatti, a friend of Marco Besso, who became a "paternal" friend of Albert in the summer of 1895 in Airolo. Their meetings took place against the backdrop of major companies in the electrical sector, with Giovanni Barberis, for example, investing in the Einsteins' business in 1895 and employing Michele Besso in 1899 when he was director of a major company, of which Marco Besso was also a member of the board of directors. These people met at international exhibitions, congresses and inaugurations. In Italy, they were also linked by the intellectual, artistic, political and revolutionary world of the *Risorgimento*, whose ramifications extended to Switzerland, a land of political asylum during the wars of independence. The meeting between Albert and Michele, whatever its date and circumstances, was in any case favoured by a predestined environment.

The possibility of a meeting organised to help Albert with his application to ETH shortly before October 1895 remains to be explored. It should be noted that Michele, who graduated in March 1895, had just left the school Albert wanted to enter in October 1895. Perhaps Michele initially acted as an informal tutor for Albert in Switzerland, before, given the six-year age gap, a real friendship developed between them later? It should be remembered that Albert himself, a year after leaving school, was the tutor of a young English student, Louis Cahen, who wished to be admitted to ETH, and that the *Collected Papers* indicate that Albert's sister Maja was the tutor of Michele's younger sister Bice in Trieste during 1902-1903, which bears witness to the customary nature of such a practice. If Michele had contributed to young Albert's training for the ETH before October 1895 and during 1896 in Aarau, where he completed his secondary education, he would not have failed to notice the originality and depth of the questions the teenager was asking himself. As we shall see in the next chapter, Albert wrote a scientific paper on the elasticity of the ether, in preparation for his admission to the ETH in October 1895. As we shall see in detail, Albert's work required knowledge that he could find scattered throughout his engineering and academic environment in Pavia, but which was also to be found in Michele's home.

The paternal branch of the Besso family

Speziali recalls that the Besso family settled in Trieste following a tragedy that befell Michele's great-grandfather, also named Giuseppe:

> Still a young man, Giuseppe travelled to Trieste on business with his second son, Salvatore [1804-1878]. On their return to Arta [a Greek town in Epirus], they learned that all the members of the family had died in the meantime from the plague. They returned to Trieste to establish themselves and trade. Salvatore acquired two boats and became a small shipowner [in the grain commerce]. In November 1836, in Trieste, he married Regina Cusin, born in 1819, whose father, Vitale Beniamino Cusin, had founded the "Assicurazioni Generali" in Trieste in 1831.

Alberto Caracciolo devotes an article to the Besso family, entitled "A diaspora from Trieste: the Bessos in the nineteenth century[81]". He focuses on their social, economic and political environment and their liberal attitude to religion. On the subject of insurance in the early 19th Century, he points out that in this city of tens of thousands of inhabitants, which was subject to a constant flow of people and goods, capital investment in the various branches of insurance, and more recently in life insurance, was among the most profitable and created links with places like Prague and Venice, Zagreb and Vienna and so on. Salvatore and Regina Besso had four sons: Giuseppe (Michele's father), Marco, Beniamino and Davide, all of whom were brought up to respect Salvatore's moral values of freedom and tolerance. Let us take a closer look at this remarkable family of siblings.

Giuseppe Besso (1839-1901), Michele's father, humanist

The eldest child in his family, he began working at an early age following his father's bankruptcy. A brilliant self-taught man, he took an interest in political economy and social issues. In 1865, according to Speziali, he became "second director of the Swiss Reinsurance Company" in Zurich. In his article, Caracciolo points out that Giuseppe showed a spirit of decisiveness and autonomy very early on, including in the ideological, political and religious spheres, and that "before moving permanently to Zurich for his business and encountering a very relaxed and modern Judaism there, he showed his widespread impatience in Trieste". He reported that Giuseppe had written a long-awaited speech on the Labour Exchange, popular credit and labour law for the Minerva Society in Trieste, one of Italy's oldest cultural associations devoted to the sciences, literature and the arts. He also reports a letter in which Giuseppe urges his brother Beniamino not to stay

---

[81] Alberto Caracciolo, "Una Diaspora di Trieste: I Besso nell'Ottocento", *Quaderni storici*, vol. 18, no. 54/3: *Ebrei in Italia*, 1983, pp. 897-912.

in Trieste: "Our future does not lie in Trieste [...] Think instead of Gotthard or southern Italy". Trieste, still under Austro-Hungarian rule, remained on the fringes of Italian unification and the patriotic and intellectual momentum that accompanied it. Davide Besso, the younger brother, complained to his brother Marco about the rather cool reception given to Italian unification by a city that profited from its trade as a maritime outlet for the Austro-Hungarian Empire, but which neglected science and literature. Giuseppe, for his part, was a fervent supporter of unification, so much so that in 1867 he proposed an international fund-raising collection to support Garibaldi's march on Rome "like the Denarius of St Peter, against him".

It should be noted that in 1864 Giuseppe wrote to the linguist Graziadio Ascoli, whom he had met earlier, telling him of the good news he had received from the family and the newspapers[82], testifying to a long-standing and deep-rooted link between the Ascoli and Besso families. Giuseppe married Erminia Cantoni in Zurich in 1872 and became a Swiss citizen in 1879, but returned to Trieste the same year, to the company's headquarters, as a member of the Board of Directors of the General Insurance company. Giuseppe died of gout in Trieste in October 1901. Speziali wrote that Michele had been deeply affected by his father's last words: "We know so little... be good!" It was a directive to which he remained committed throughout his life.

Beniamino Besso (1840-1907), railway engineer and scientific author

After studying in Trieste at the Imperial Royal Academy of Commerce and Shipping, Beniamino enrolled at the Faculty of Mathematics of the university of Padua in 1857 and obtained his doctorate in Mathematics in April 1861[83]. He appeared in Turin in 1869 as a non-resident member of the Society of Industrial Engineers. The following year, he was an engineer with the Upper-Italian Railways design office and in 1872 he was in Florence, as the Turin register still states. He then became director of the Sardinian Railways in Rome, a company owned by English and Italian industrialists and politicians. As we saw in the first chapter, the development of the railways played a strategic role in the new Italy, and Beniamino played a major political role, working with members of Parliament, ministers and industrialists on the route and financing of new railway lines, for example.

He was also the author of a number of books, as illustrated above by the book read by Michele as a child. In 1870, Beniamino published *Railways*[84], which not only gave a historical overview of the development of the railways,

---

[82] Susanna Panetta, *Il diligentissimo inventario dell'archivio di Graziadio Isaia Ascoli*, 2014.
[83] Communication from the archivistic office of the university of Padua.
[84] Beniamino Besso, *Le strade ferrate*, Milan, Treves, 1870.

but also detailed the methods used to lay out the lines and build the tracks, tunnels and bridges, as well as the characteristics of the locomotives and the Italian and European networks. In 1871, he published a small booklet entitled *Mont Cenis, illustration and description*[85], which presented the technical realisation of the Fréjus tunnel and was published for its inauguration on September 17, 1871. Galileo Ferraris, whom we met as an illustrious scientist in the field of electrical engineering, and Beniamino Besso probably crossed paths. While studying at the Museo Industriale in Turin, Ferraris visited the Fréjus tunnel construction site in 1868 and in 1870 attended the banquet in honour of the engineers who had completed the tunnel[86], organised by the Society of Engineers, of which Beniamino was a member and which Galileo joined in 1873. These events bear witness to the intertwined destinies of railway and electrical engineers. So, it's not surprising that Beniamino also wrote a book on *Electricity and its Applications*[87]. In it, he gives a richly documented educational account, based on the history of science and technology, right up to the most recent applications in the field. The book begins with a chapter on electrostatic machines, then moves on to the lightning rod, accompanied by a long biography of Benjamin Franklin, the Volta battery and the controversy with Galvani, electric dynamos (including the Alliance machine mentioned in the previous chapter), aerial and underwater telegraphy and even electroplating. In a similar vein, Beniamino had also published by Treves *Steamships and lighthouses* in 1869[88], in which he explained how steamships worked, placing them in their historical and technical context, and described how Augustin Fresnel had improved lighthouses. As we have seen from his work on the great discoveries, Beniamino also belonged to the circle of scientists who took part in the movement to popularise science. This book was reprinted six times between 1864 and 1873. Beniamino's distinguishing feature was his concern to place knowledge in context and to give a historical account of it. Michele's repeated appeals to Albert to write a detailed history of the genesis of his scientific ideas, which contributed to the history of mankind, as we saw in the introduction, probably originated in Beniamino's particular sensibility, which Michele had encountered from his earliest childhood and then in Rome during his adolescence.

Like his brother Marco, Beniamino was linked to Pietro Blaserna by a shared interest in scientific dissemination – of which Blaserna was also one of the main architects in Italy. The railways worked with physicists from the Institute of Physics in Rome, founded by Blaserna, to solve technical problems relating to the electrification of the lines. They came from the same region, Blaserna from Gorizia near Trieste, and shared the same political sensibilities. Like his brothers, Beniamino remained committed to Trieste's

---

[85] Beniamino Besso, *Il Cenisio, illustrato e descritto*, Torino, Mattirolo, 1871.
[86] Raffaella Gobbo, "L'archivio di Galileo Ferraris", *Rassegna degli archivi di stato*, nuova serie, anno 1, no. 1-2, 2005, pp. 9-169.
[87] Beniamino Besso, *L'elettricità e le sue applicazioni*, Milan, Treves, 1871.
[88] Beniamino Besso, *I battelli a vapore ed i fari*, Milan, Treves, 1869.

independence. Together with Blaserna, he helped finance the studies of the activist Guglielmo Oberdan (1858-1882) in his second year of engineering school in Rome[89]. Marco Panconesi, in his biography of Italian railway engineers, says of Beniamino: "During his stay in Rome in 1879, he met the young irredentist patriot Guglielmo Oberdan, also from Trieste, whom he tried to help by getting him a job with the Sardinian Railways as a technical draughtsman for locomotives".

Beniamino's wife, Amalia Goldmann Besso (1856-1929), also born in Trieste, was a well-known painter, to whom the Italian online encyclopaedia *Treccani* devotes an article. It states that she moved to Rome with Beniamino in 1883, when the couple married. An exhibition entitled "Artists of the 19th Century through Jewish vision and identity" was dedicated to her by the city in 2014. The paintings, on loan from the Marco Besso Foundation, illustrate Amalia's trip around the world (to Russia and Japan) in 1912 with her nephew, Salvatore, Marco's son, who died shortly after his return. The painter Liliah Nathan (1868-1930), wife of Moisè Ascoli and one of the seven daughters of the aforementioned Ernesto Nathan, was also exhibited here, along with her sister Anna. Anna, Liliah and Amalia played an active role in the emancipation of women in Italy at the beginning of the 20th Century.

Marco Besso (1843-1920), the influential Chairman of General Insurance

Marco Besso, the second son, had to turn away from university at the age of sixteen and, like his elder brother, enter working life by taking a job as a trainee at General Insurance. He quickly rose through the ranks. In his first job in Milan in 1866, at the age of twenty-three, he replaced the head of one of the company's offices, who had left to enlist in Garibaldi's troops. Marco explains in his autobiography[90] that he was inducted to the Insubria Masonic lodge, which included lawyers, bankers and academics from Milan. The lodge takes its name from the people who are said to have founded the city of Milan and was founded in 1864 as a result of a split with the Grand Orient of Italy on the basis of a more social political programme, under the leadership of Ausanio Franchi[91]. In 1877, Marco Besso was appointed

---

[89] Oberdan, accused by the Austrians of plotting an assassination attempt on Emperor Franz Joseph, was sentenced to death and hanged in 1882; see Eva Cecchinato and Daniele Ceschin, "Guglielmo Oberdan", in *Dizionario biografico degli Italiani*, vol. 79, 2013.

[90] Marco Besso, *Autobiografia*, Rome, Fondazione Marco Besso Editrice, 1925, repr. 1970.

[91] "In the speech he [Ausanio Franchi] read on May 30, 1864, he outlined a vast project in which he called for the creation of mutual aid societies and credit institutes for artisans and peasants; he asked for schools open to all and access to work, instead

General Secretary of the Central Committee of General Insurance Directors. As already mentioned, he also chaired the board of directors in Milan of the accident insurance company set up by the company in 1896. He was also a member of the London Society of Actuaries, which brought together insurance company statisticians keen to improve their practices. In 1909, Marco was elected Chairman of General Insurance by its Board of Directors. At the time, General Insurance was Italy's largest insurance company and one of the largest in the world. In 1876, he published an essay on the operation of pension funds, using the example of the Upper-Italian Railways, where his brother Beniamino worked, and then took part in the rail fares commission set up by the Government. He was also involved in financing the electricity industry: in 1891, he raised the capital needed to set up the Veneto Telephone and Electricity Company, which challenged the monopoly of the local company for electric lighting[92] and became a member of the board of directors of the Society for the Development of Italian Electrical Companies in 1899, which at the time employed his nephew Michele, a company of which he became Chairman in 1914[93].

Marco Besso was also elected a corresponding member of the Accademia dei Lincei in 1920 for his literary works. In 1904 he published *Rome and the Pope – through proverbs and sayings*, in 1912 *The Fate of Dante outside Italy*, in 1914 *Philobiblon* "inspired by Riccardo de Bury and dedicated to his native Trieste" and in 1918 the *Encomium Moriae of Erasmus of Rotterdam*[94]. Its foundation, set up in Rome in 1918, houses a rich library devoted to the various editions of the works of Dante Alighieri and Erasmus of Rotterdam, as well as studies on these authors, which are still a valuable source for researchers today[95]. It also includes a number of original documents on the Lazio region, placing the new Italian capital in its historical and millenary perspective. This testifies to Marco Besso's attachment to the city and his support for the project to make Rome the capital of a unified country. Michele went to Rome in 1919 to classify his uncle's library.

In 1874 Marco Besso married Ernesta Pesaro Maurogonato (1853-1916), daughter of Isacco Pesaro Maurogonato and Bersabea Ascoli, in Florence,

---

of charity and benevolence. From a political point of view, he felt that Freemasonry should be a neutral field where everyone could meet for the common good. His Masonic ideal can be summed up as follows: to reduce humanity to a single family" (Anna Maria Isastia, "La Massoneria", in *L'Unificazione*, 2011, online at www.treccani.it).

[92] Giorgio Mori (Ed.), *Storia dell'industria elettrica in Italia*, vol. 1, *Le origini, 1882-1914*, *op. cit.*, p. 269.

[93] Arianna Scolari Sellerio Jesurum, "Marco Besso", in *Dizionario biografico degli Italiani*, vol. 9, 1967. This company has played an important role in mediating conflicts between Italian and foreign industrial groups over the sharing of their spheres of influence. It is 20% financed by Milanese bodies and the rest by an Austro-German group.

[94] Book titles translated from Italian.

[95] Marco Besso Foundation, Largo di Torre Argentina 11, 00186, Rome.

where he lived until 1876. The latter was the sister of the aforementioned Graziadio Ascoli. Two of their four children died in infancy. Marco Besso's correspondence includes several letters from Graziadio between 1883 and 1902 in which he calls Marco "my dearest nephew". The affinity between the uncle from Gorizia and the nephew from Trieste extended to politics. In July 1895, Marco sent Ascoli his congratulations on the publication of his article on the irredentists, in which he advocated the armed annexation by Italy of Julian Veneto (a region bordering Slovenia, Croatia and Italy) at the expense of the Austrians. His attachment to his home region was also evident in the three annual university scholarships awarded by his foundation between 1924 and 1936 to young people from Trieste, in memory of his son Salvatore (1884-1912). The Besso family's openness to Central Europe certainly also influenced Michele, who looked after Albert's family, in particular Mileva, who was of Serbian origin, from 1915 onwards. Finally, Marco Besso co-signed the manifesto for the creation of the Dante Alighieri Society with 170 other personalities, including his uncle by marriage Graziadio Ascoli and Ernesto Nathan (1845-1921). Nathan, Grand Master of the Grand Orient of Italy from 1896 to 1904 and Mayor of Rome from 1907 to 1913, was also the father-in-law of the physicist Moisè Ascoli, Graziadio's son, whom we will mention again in connection with Michele Besso's thesis. The Society's manifesto, which aimed to promote the Italian language and culture abroad, was published in Rome in July 1889.

Marco's correspondence, which is held in his foundation in Rome, shows that the physicist Pietro Blaserna knew the Besso family independently of the fact that Michele had been his student. It includes a visiting card on which Blaserna wrote a friendly note intended to introduce him to a third person, beginning "My dear friend". The finance minister Luigi Luzzatti also used these words in the hundred or so letters he sent him. There are also a few letters from Salvatore Majorana, Quirino's father, when he was Minister of Agriculture, Trade and Industry. In these letters, dated May 1879, which concern the creation of mutual aid societies for workers, the minister thanks Marco for his support for his project. There is also an abundance of correspondence with leading European figures, testifying to the fact that Marco Besso was a very important Italian figure.

Davide Besso (1845-1906), mathematician and teacher

Davide, the younger brother, studied in Trieste at the Royal Imperial Academy of Commerce and Navigation and at the Hydrographic Institute from 1857 to 1861, before going on to study at the University of Pavia. The university's historical archives have preserved his student record. In 1862, Davide wrote to the rector, the physicist Giovanni Cantoni, to enrol initially as an auditor in the mathematics section of the Faculty of Physical, Mathematical and Natural Sciences. He explained that, as a native of Trieste, he would have liked to be able to matriculate at the University of Padua, which would have been natural given the similarity of the curricula between that university and his school. But the Austrian Government would not allow him to do so and would only allow Venetian students to enrol. Davide certainly found a sympathetic ear in Cantoni, who had gone into exile in Lugano in 1849 after the Austrians returned to Milan. He undertook to sit the entrance exam to the University of Pavia to regularise an unofficial enrolment and passed with flying colours, scoring 30/30. He was taught mathematics by Felice Casorati and physics by Cantoni and obtained his degree in pure mathematics on July 23, 1865 with 30/30 in complementary algebra, differential and integral calculus, descriptive geometry and rational mechanics and 28/30 in physics. He then graduated in mathematics from the University of Pisa in 1866, where he studied theoretical geodesics under Eugenio Beltrami, who would become known for his contribution to non-Euclidean geometries, as well as under Enrico Betti, a friend of Riemann's, known in particular for his work on topology. Appointed professor of mathematics at the Leonardo da Vinci Technical Institute in Rome in 1871, he became a professor at the University of Modena in 1888, where he taught algebra. We should point out that he was then a colleague of Roberto Schiff, who taught chemistry there and who is quoted in Albert's first published article, on molecular forces and capillarity, one of the topics discussed with Michele in Milan in 1900, to which we shall return in Chapter 6.

In 1886, Davide Besso founded the *Periodico di Matematica*, a journal for mathematics teachers, of which he was Editor-in-Chief until 1896, when he took early retirement after being promoted to full professor. The journal promoted the renewal of mathematics teaching in secondary schools, as did other contemporary European journals. In 1905, for example, it published in Italian Henri Poincaré's lecture entitled "General definitions in mathematics", delivered at the Pedagogical Museum in Paris in 1904, as part of the Georges Leygues reform of 1902 in favour of scientific education. The journal became *Periodico di Matematiche* in 1921 with Federigo Enriques, who in 1922 presented Einstein's relativistic theories. As far as his research was concerned, Davide Besso was a specialist in fifth- and sixth-degree differential equations, continuing the work of Brioschi that had been noted by Felice Casorati. His ten-page obituary, published in *Periodico* in 1907, lists some sixty publications, most of which appear in the proceedings of the Accademia dei Lincei in Rome[96]. Roberto Marcolongo also tells us that in retirement Davide devoted his time to his mathematical library of rare works, a passion similar to that of his brother Marco. He emphasises the rigour and care that Davide Besso brought to his lessons, as well as the straightforwardness of his character, an essential trait of which was modesty: "he was modest, incredibly modest". A trait that Michele also possessed, alongside his remarkable mathematical abilities.

---

[96] Roberto Marcolongo, "Davide Besso", *Periodico di Matematica* **22**, 147-156 (1907).

The maternal branch of the Cantoni family

Pierre Speziali, in his introduction to the Einstein-Besso correspondence, gives no information on this maternal branch of the family, except that it originated in Mantua. Michele Besso's mother, Erminia Cantoni (1852-1922), was the daughter of Angelo Cantoni (1825-1887), a wealthy Milanese banker and Elisa Maroni (1829-1890). Angelo's capital was officially valued at 500 000 lire in his share of the bank Angelo Cantoni & Cie, 302 000 lire in securities and 105 000 lire in land[97]. He invested 150,000 lire in Filatura Cantoni when the latter, managed by Eugenio Cantoni, increased its capital to become the first Italian company to be listed on the stock exchange. Despite sharing the same surname, Eugenio and Angelo's families were probably not related, except professionally. Erminia had two brothers, Vittorio and Tullo, and three sisters, Bice, Maria and Emma.

Maria (1855-1940) married her cousin Achille Cantoni (1848-1914). In 1922, Albert received Arrigo, one of their five children, as Michele tells us: "The last I heard of you reached me through my cousin Arrigo Cantoni, who was very pleased about your friendly welcome". In 1905, Arrigo (1877-1953) graduated in architecture from Turin Polytechnic. It was probably he who presented a project in 1912 for Milan's new railway station. After the First World War, he became a teacher of experimental physics at secondary school and in 1938 published a book of physics experiments with the Milanese publisher Hoepli.

Tullo Cantoni, born on September 27, 1866, in Vicenza, was Erminia's younger brother. He completed his third year of law studies at the University of Bologna in 1888, after two years at the University of Rome. After converting to Catholicism, he obtained the title of Count Mamiani della Rovere. Mayor of Arona, a town in northern Piedmont, from 1914 to 1917, he lived in the Villa Cantoni on the shores of Lake Maggiore with his wife Irma Finzi and their children. His son Vittorio Angelo, born in 1899, and Irma disappeared on September 15, 1943 during a Nazi roundup, as commemorated by a recently inaugurated stele in the town. Her twin sister, Emma (1866-1946), married Eugenio Norsa (1856-1933) on July 5, 1888, a descendant of one of Italy's oldest and most famous Jewish families, who imported silk from Japan for the textile industry[98].

Below we devote specific paragraphs to two of Michele's uncles whom we have already met: Vittorio Cantoni, a civil engineer who graduated from

---

[97] Germano Maifreda, *Gli Ebrei e l'economia milanese: l'Ottocento*, Milan, FrancoAngeli, 2000. Davide Maroni (Elisa's father or uncle?) invested 75,000 lire in Eugenio Cantoni's business.

[98] The history of this family was retraced in 1950 by Paolo Norsa in his book *Una famiglia di banchieri, la famiglia Norsa (1350-1950)*.

ETH in Zurich, and Giuseppe Jung (1845-1926), a professor of mathematics at the Milan Polytechnic, who married Bice Cantoni in 1879.

Finally, let us note that Michele Besso's brothers and sisters were named after their grandparents, uncles and aunts, as was customary at the time: Salvatore (1876-1968), Vittorio Beniamino (1886-1937), Marco Tullo (1880-1898) and Bice Luisa Margherita (1890-1965).

Vittorio Cantoni (1857-1930), civil Polytechnic engineer

Vittorio Cantoni is well known for his contribution to the electrification of Italy; however, he had not been linked to Michele's family until our study and it is important to provide biographical details about him. In 1874, at the age of seventeen, he enrolled at ETH Zurich, graduating on March 23, 1878. His student file states that Michele's father, Giuseppe, is his "correspondent" in Switzerland, and Vittorio gives his address in Zurich. He therefore lived with Michele's parents during his early years of study, and, when he left Zurich, Michele was five years old. At ETH, Vittorio entered the mechanical engineering section, but switched to civil engineering the following year. It should be noted that he had passed Weber's physics exams with top marks in his second year of study in 1875-1876[99]. Vittorio returned to Italy and graduated from the Milan Polytechnic on September 9, 1879. Vittorio's background was probably not unrelated to Michele's decision to study at ETH in 1891 after his first year of undergraduate studies in Rome.

Vittorio was soon entrusted with important responsibilities. As we have seen, in 1886 he helped install Gaulard-Gibbs transformers at the Tivoli hydroelectric power station, 28 km from Rome. This installation used electric dynamos from the Ganz company in Budapest. The power station is located halfway up the Tivoli waterfall, with a drop-height of 50 metres. A note from Lacaita states: "Cantoni was the delegate adviser of the Company for hydraulic forces for industrial and agricultural use and entered into an agreement with Gaulard to install the city's lighting using the new distribution system[100]". He later became Chief Engineer of the Tivoli-Rome power line, inaugurated on July 4, 1892 in front of 500 VIPs. The 28 km line was the first in Italy to carry alternating current. It was designed to carry the electricity generated in Tivoli to Rome to light the city[101]. Vittorio was necessarily directly linked to Galileo Ferraris, if only for the installation of the

---

[99] Vittorio Cantoni, Anmeldung zur Aufnahme in das Eidgenössische Polytechnikum, ETH Zurich Archives.

[100] Carlo G. Lacaita, "Politecnici, ingegneri e industria elettrica", in Giorgio Mori (Ed.), *Storia dell'industria elettrica in Italia*, vol. 1, *Le origini, 1882-1914, op. cit.*, p. 642, note 44.

[101] P. Marcillac, "Le transport d'énergie électrique de Tivoli à Rome", *La Lumière électrique* **40**, 7-16 (1893).

transformers, for which the latter had carried out the theoretical study in 1885, as we saw in the previous chapter. It is likely that he was also present at the International Electrical Exhibition in Frankfurt in 1891, along with Ferraris, where his former teacher, Weber, was in charge of the testing committee for the first alternating current transmission line.

Vittorio also oversaw the construction of the Cantoni family's sumptuous villa on the shores of Lake Maggiore, located beneath the rock of Arona. The land belonged to the aristocratic Borromeo family. A recent study published by the University of Turin[102], devoted to the villas and gardens on the shores of Lake Maggiore, mentions around one hundred and fifty villas, but only five are the subject of detailed articles, including Villa Cantoni. The author of the article, Paolo Cornaglia, dates its construction to between 1883 and 1887. The villa features numerous frescoes painted by Giacomo Casa (1827-1887). A lush garden of exotic plants overlooks the lake, where a pier awaits residents. The Marco Besso Foundation has a photograph of the Cantoni-Besso families together at the villa. It is highly likely that Michele stayed there and it would not be surprising if Albert, when he climbed a mountain near Lake Maggiore in September 1900, for example, had been invited. Today, the property is abandoned[103] and no longer belongs to the Cantoni family.

There is one astonishing detail that gives us an unexpected insight into the Cantoni family's relationship with the Italian intellectual elite. In June 1872, for the wedding of his sister Erminia, Michele's mother, to Giuseppe Besso, Vittorio, who was barely fifteen, had a thirty-two-page booklet published entitled *In praise of Salomon Gessner* by Filippo Mordani (1797-1886). Vittorio had undoubtedly read the Romantic works of Gessner (1730-1788), a writer in vogue at the time who extolled the charms of Zurich country life. However, he justified his dedication by saying that the praise seemed very appropriate to the circumstances – his sister's departure for Zurich – adding that the text "was written by a remarkable writer from Ravenna, F. Mordani, who was generous with the request made to him by one of his friends so that we could use it". The text in question had been read at the prize-giving ceremony of the Ravenna Academy of Fine Arts in 1840 and Mordani had been elected honorary member the following year. Gessner's works were translated into Italian by the poet Andrea Maffei, husband of the Countess Clara Maffei we met in the first chapter[104]. Mordani's eulogy of Gessner explicitly pays tribute to Maffei's translation of the *Idylls*. How did such a specific piece of writing come to the attention of a young man of fifteen? Who was

---

[102] Renata Lodari, *Giardini e ville del lago Maggiore. Un paesaggio culturale tra Ottocento e Novecento*, Turin, Centro Studi Piemontesi, 2002.

[103] It remained in a state of neglect for over thirty years until renovation work began in 2016 – source Wikipedia.

[104] Earlier, we mentioned the lighting of the Maffei locomotive company in Munich by J. Einstein & Cie. This Munich family may be related to the one in Milan: in 1816, Giuseppe Maffei, Andrea's uncle, also moved to Munich, where Andrea visited him.

Mordani's friend who authorised its reproduction by Vittorio? Maffei himself? Or Count Terenzo Mamiani della Rovere (1799-1885), a major figure in the *Risorgimento*, professor of philosophy and Cavour's minister, who had offered Mordani a post in Genoa? We should remember that Tullo Cantoni, Vittorio's younger brother, inherited the title of Count Mamiani della Rovere.

Giuseppe Jung (1845-1926), professor of graphic statics at the Milan Polytechnic

Giuseppe Jung married Bice Cantoni, Michele's aunt, in 1879. He began his career as a teacher at the Lyceum Parini in Milan (formerly the Royal Lyceum of Brera), where he also taught a course for students wishing to enter the Polytechnic without having to complete the first two years at the mathematics faculty of the University of Pavia. One of eight siblings, he is an uncle of Italian Finance Minister Guido Jung, to whom two recent biographies have been dedicated[105]. From the start of the 1873 academic year, Giuseppe occupied the chair vacated by Luigi Cremona at the Polytechnic, when Cremona was appointed director of the School of Engineering in Rome by Francesco Brioschi and Pietro Blaserna. He was in charge of courses in Projective Geometry and "Graphic Statics" from 1877 to 1913. This discipline, pioneered by Luigi Cremona, is a branch of geometry that uses graphical methods to solve mechanical problems by analysing forces in equilibrium. In 1885, Jung wrote the introduction to Cremona's book *Reciprocal Figures in Graphic Statics*. He became a correspondent of the Lombard Institute in 1879, a permanent member in 1893 and a resident in 1908. He died in January 1926, and a tribute to him was published in 1927 by Gian Antonio Maggi in the Annals of the Institute, with a full list of his publications[106].

Albert Einstein knew Giuseppe Jung personally and had told Mileva that he was one of his "good connections" in Milan, as we have already mentioned. In his letter of April 4, 1901 already commented on, he added: "The day before yesterday he [Michele] went on my behalf to see his uncle Prof. Jung, one of the most influential professors of Italy & also gave him our paper. I met the man once before & must admit that he impressed me as quite an insignificant person. He promised that he will write to the most important professors of Italy". We shall see that the memoir in question was Albert's article on capillarity, which had just been published in the *Annalen der Physik* and which Jung was to give to Professors Righi and Battelli. As for the first meeting, it probably took place in Zurich in 1897, on the occasion of

---

[105] Nicola De Ianni, *Il ministro soldato. Vita di Guido Jung*, Naples, Rubbettino, 2009; Roberta Raspagliesi, *Guido Jung. Imprenditore ebreo e ministro fascista*, Milan, FrancoAngeli, 2012.
[106] Gian Antonio Maggi, "Giuseppe Jung, Commemorazione letta nell'adunanza del 14 aprile 1927", *Rendiconti Istituto lombardo* 60, 2 (1927), pp. 291-307.

the first international mathematics symposium[107]. The colloquium was organised at ETH by Albert and Michele's professors (Carl Friedrich Geiser, Adolf Hurwitz, Ferdinand Rudio, Heinrich Friedrich Weber and Albin Herzog) and it would be surprising if the school's mathematics and physics classes had not been invited to attend. The colloquium was held under the auspices of Henri Poincaré for France and Félix Klein for Germany[108]. Michele Besso and Giuseppe Jung appear on the list of participants, and Jung's address is given as 9 via Borgonuovo in Milan, two hundred metres from the Einstein family residence and on the way to the Lombard Institute. Chapters 5 and 6 look in detail at the contents of Jung's personal library, which bears witness, among other things, to his ongoing relationship with Albert until 1906. Jung was also involved in Einstein's appointment to the Institute in 1922.

[107] http://www.mathunion.org/ICM/.

[108] Anne-Marie Décaillot, *Cantor et la France. Correspondance du mathématicien allemand avec les Français à la fin du XIXe siècle*, Paris, Kimé, 2008. Poincaré, Klein and Hurwitz were each asked to give a plenary lecture. Poincaré's paper "On the relationship between pure analysis and mathematical physics" was read in his absence.

4
Albert's environment in Pavia and preparations for ETH

In December 1894, Albert Einstein left his Munich grammar school to join his family in Milan. He spent part of 1895, until September, in Pavia, some forty kilometres south of Milan, where the family's production plant was located. He stayed there occasionally until 1900. In Pavia, he worked in his uncle Jakob's engineering office at the firm Einstein, Garrone & Cie. He grew up in an environment of Polytechnic engineers where, as we have seen, he familiarised himself with electrotechnical issues and was able to consult journals in the field. This technical environment and the confidence his family placed in him led him to apply for admission to one of Europe's most prestigious schools, the Federal Institute of Technology in Zurich (ETH), in October 1895, even though he was not old enough to sit the entrance exam and had not completed his secondary education: he was only sixteen and had not attended school in Italy.

We shall see how the plan to enter the school mobilised the whole family from August onwards: Albert wrote personally to the illustrious professor of electrical engineering Galileo Ferraris, in an attempt to obtain a letter of recommendation from him to Heinrich Friedrich Weber, the professor of physics and electrical engineering at ETH; his parents sought the intervention of a Swiss banker and family friend, Gustav Maier, from the Rector of the school, Albin Herzog, so that he would be allowed to sit the exam. Another way of overcoming these difficulties was to write an early scientific paper, shortly before October 1895, on the state of the ether in a magnetic field. This dissertation dealt with Herzog's and Weber's teaching areas, and with the problem of the nature of electric current, which must have concerned Albert very early on. Through his partners in the firm Einstein, Garrone & Cie, Albert was also immersed in an academic environment that could have been useful to him in his revision of the ETH curriculum. More generally, it is likely that the scientific environment in Pavia gave him a very early awareness of the issues to which he would later return: the phenomena of capillarity, the subject of his first paper submitted in December 1900, in which the physicist Carlo Marangoni, the uncle of his friend Ernestina in Pavia, was a renowned specialist, and perhaps also Brownian motion, the kinetic explanation of which had been provided by the famous Pavian physicist Giovanni Cantoni, whose student and assistant Carlo had been.

Albert's attempt to be admitted to ETH in October 1895, at the age of sixteen without a school-leaving certificate

The plan to be admitted to ETH

When his family was in Pavia, Albert worked in his uncle Jakob's engineering office, as Otto Neustätter attests in a letter to him on his fiftieth birthday in 1929. He recalls the pride with which Jakob spoke of him at the time: "You know, my nephew is already fabulous. Where I and my assistant engineer had been racking our brains for days, the young fellow figured out the whole thing in hardly a quarter of an hour. I tell you, he's going to do great things![109]". Jakob recognised Albert's talents very early on and his marked taste for mathematics, a subject in which he had trained him from the age of twelve. Albert was familiar with the world of Polytechnic engineers, and it was only natural that he should consider going to a Polytechnic, just like his uncle. His mother tongue is German, but he has no intention of returning to Germany to study, so it's only natural that he should consider the ETH in Zurich, which is also one of the most renowned schools in Europe. A further reason for this choice was that Weber, the school's professor of electrical engineering, was a renowned scientist who was well-known in the family circle.

The examination syllabus includes a general part, detailed on one page in the student booklet. Candidates are required to write a table-top essay in one of the four languages in which the courses are taught (German, French, Italian and English) and then take an oral examination to show that they have a sufficient command of these languages to follow the courses. They then sit an oral exam in three parts, covering literature, the political history of Switzerland and the natural sciences. The scientific part is the most important. The written exam focuses on mathematics, the syllabus for which is detailed on two pages. It covers: arithmetic operations on whole and fractional numbers, first and second-degree equations with one or more unknowns, logarithms, exponentials, arithmetic and geometric sequences and development in series, elementary combinatorics; plane and spatial geometry,

---

[109] Letter from Otto Neustätter in *CPAE (English translation supplement)*, vol. 16, Doc. 430, p. 390.

calculation of volumes and surfaces of solids, trigonometry and elements of spherical trigonometry, analytical geometry, conics, and finally representative geometry. Physics is mentioned in a short paragraph, but in relation to topics covering a broad spectrum: elements of point and fluid mechanics, sound waves, elements of heat theory (thermodynamics), geometric optics, the main properties of optical and thermal radiation, magnetic and electrical phenomena and their laws.

The letter to Galileo Ferraris

In August 1895, Albert was still unsure whether he would be able to sit the exam. He had the idea, probably suggested by his uncle Jakob, of writing a letter to Galileo Ferraris, a friend of Weber. His letter was recently discovered by Andrea Silvestri, professor of electrical engineering at the Polytechnic of Milan and member of the Lombard Institute, responsible for the Ferraris archives[110]. Dated August 12, 1895, it was between the two pages of a book in the scientist's personal library and is the oldest letter written by Albert to have been found to date. Written on the family company letterhead, it bears the address of the Einsteins' first residence in Milan, via Berchet 2, which was also their company office before they moved to Pavia. It is signed "Albert Einstein, from Ing. Einstein, Garrone & Cie, Pavia". The young Albert thus presents himself as a fully-fledged member of the family firm's engineering office, which corroborates Neustätter's earlier remark. Albert begins by thanking Ferraris for remembering him, based on what he says the engineer Vitali had told him on his return from Turin. We can deduct from this that he had been introduced to Ferraris earlier, probably when he was only twelve, at the Frankfurt exhibition of 1891, according to Chapter 2. He informed Ferraris of his intention to study at ETH from the beginning of the 1895 academic year, and of the difficulties he had encountered because of his young age. He therefore asked him for a "short private recommendation" to Professor Weber so that he could attend his classes. Maja Einstein wrote in her brother's biography that Weber had indeed authorised him to do so, which she attributed to her brother's brilliant marks in scientific subjects in the entrance examination in October 1895. According to her, he had failed because of poor marks in languages and history[111]. For Carl Seelig, it was languages and natural sciences that caused him to fail, despite his remarkable results in mathematics and physics. It should be noted, however, that there is no record of Albert taking a written exam, and that his audition could have been less formal. Albert himself states in his *Autobiographical Sketch* that "it was nevertheless comforting that the physicist H. F. Weber had told me that

---

[110] Andrea Silvestri, *Omaggio ad Albert Einstein*, Milan, Politecnico di Milano, 2005.
[111] "Albert Einstein - A Biographical Sketch" (Translated Excerpts) by Maja Winteler-Einstein in *CPAE (English translation supplement)*, vol. 1, pp. xv-xxii.

I could listen to his colleagues if I stayed in Zurich[112]" and that an interview on the general knowledge part had revealed his shortcomings.

Intervention with Herzog

During the summer of 1895, the family tried to overcome the administrative difficulties associated with Albert's young age and his lack of a degree. They contacted a friend of Hermann Einstein's, Gustav Maier (1844-1923), whom he had met at Ulm. Maier had temporarily retired from business to live in Switzerland, in the canton of Thurgau, after managing the Reichsbank in Frankfurt. He had recently moved to Zurich[113]. Maier pleaded Albert's case by writing to the Rector of ETH, Albin Herzog. Although the text of his letter is not known, Herzog's reply, dated September 25, 1895, is published in the *Collected Papers*. Herzog indicated that it was not *a priori* desirable to remove a student from the institution that had trained him, even if he were a "child prodigy" – his use of inverted commas shows that he was clearly borrowing Maier's own terms. Herzog advises him to convince Albert's parents to let their son complete his secondary education and take his Matura. He left the door open, however, specifying that an exceptional dispensation could be granted if the parents disagreed (which was clearly the case), but he added the following express condition: "that the Rector of the educational establishment in question is going to confirm in writing and to the fullest extent your information regarding the talents and intellectual maturity of the candidate[114]". However, Albert was no longer at school in Italy and all he had in his possession was a simple certificate from his mathematics teacher in Munich, dated December 1894, which could certainly not be used as an administrative document for a Swiss school. So how could Herzog be convinced of his talents and intellectual maturity on a scientific basis?

We tender the hypothesis that the paper written by Albert entitled "On the Investigation of the State of the Ether in a Magnetic Field" was intended for Herzog and to prove that Albert had the knowledge required for the programme (and even more). In fact, the article refers to elasticity (Herzog's own field of teaching) by discussing the speed of waves as a function of the

---

[112] Albert Einstein, "Autobiographische Skizze", in Carl Seelig, *Helle Zeit – Dunkle Zeit*, Zurich, Europa Verlag, 1956, pp. 9-17. Translation from German.

[113] Carl Seelig states on the very first page of his biography of Einstein in Switzerland that Maier had been "appointed general manager of the Brann stores and of a weaving factory. He spent his spare time in writing, and until his death in 1923, wrote with idealism and foresight on national economic, social and pacifist themes. He was one of the founders of the Swiss section of the 'German Society for Ethical Culture' [which he chaired from 1896], was a member of a Freemason Lodge and took endless pains to instil into youth a spirit of gentility, a serious sense of responsibility and understanding for international peace".

[114] Letter from Albin Herzog to Gustav Maier in *CPAE (English translation supplement)*, vol. 1, Doc. 7, p. 7.

density of the medium in which they propagate and the restoring forces that occur. He also draws on magnetic and electrical phenomena, as well as optical and thermal radiation (Weber's domains) by evoking the laws of induction, magnets and electric currents, and by looking at the speed of propagation and polarisation of electromagnetic waves in the ether. But it goes beyond the scope of the ETH entry programme, since, for example, elasticity and propagation concern an anisotropic medium, the magnetostriction effect (internal tension of magnetised bodies) is applied to the ether, and the article is presented from the perspective of experimental research for a better understanding of the nature of electric currents. We return to this unusual memoir in some detail, before turning to Albert's entourage in Pavia and the links he enjoyed with the academic world through his uncle's associates or his connection with Carlo Marangoni, the uncle of his friend Ernestina.

The scientific memoir of 1895

The circumstances

The memoir entitled "On the Investigation of the State of the Ether in a Magnetic Field[115]", written by Albert in the summer of 1895 at the age of sixteen, is likely to have played an important role in the young man's life, if only because it brought him face-to-face with the demanding criteria of scientific writing at a very early age. This astonishing document appears on the very first pages of the first volume of the *Collected Papers*, of which it is the fifth in chronological order, after Albert's birth certificate, a letter from his mother to his sister Fanny, and two short comments by Albert, one concerning a mathematical demonstration and the other a philosophical discussion. It is written on three sheets, probably dated summer 1895. It begins with a warning to the reader:

> The following note is the first modest expression of a few simple thoughts on this difficult topic. It is with reluctance that I am compressing them into an essay that resembles more a program than a treatise. As I was completely lacking in materials that would have enabled me to delve into the subject more deeply than by merely meditating about it, I beg you not to interpret this circumstance as a mark of superficiality. May the indulgence of the sympathetic reader match the humble feelings with which I present these lines.

If the first sentence shows that Albert is interested in a subject that seems topical to him, the second attests that he feels somewhat obliged to write it, and the last two indicate that this work is intended for a competent person who must judge its value. Given the circumstances, it's hard not to think of

---

[115] "On the Investigation of the State of the Ether in a Magnetic Field" in *CPAE (English translation supplement)*, vol. 1, Doc. 5, pp. 4-6.

an ETH professor. This article (probably a copy) was sent for information to his favourite uncle, Caesar Koch, a brother of his mother Pauline, who had lived in Zurich. In the letter accompanying the paper, sent from Pavia, Albert warned him that it "deals with a very special topic", even telling him that he would not be offended if he did not read it and adding in conclusion:

> As you already know, I should now enter the Polytechnikum in Zurich. This matter encounters considerable difficulties because I should be at least two years older for it. We shall write you in the next letter about the outcome[116].

It would therefore appear that the memoir was indeed linked to the difficulties mentioned and that Albert expected it to play a decisive role. As the letter is undated, its content, reproduced above, makes it possible to place it a little before the ETH examination, which took place at the beginning of October 1895, though without any further details.

The subject: the state of the ether in a magnetic field

In his dissertation, Albert looked at the effect of a static magnetic field, whatever its origin (currents, magnets), on the speed of propagation of electromagnetic waves in the ether. At the time, no one doubted the existence of this hypothetical medium. Albert qualitatively linked the establishment of an electric current in a wire, still a mysterious subject, to a transient setting in motion of the surrounding ether; he indicated that in the stationary state, when the current is constant, the deformation forces undergone by the elastic ether cause a tension in the medium and a potential energy that manifests itself in the presence of a magnetic field. The latter must therefore have an effect on the speed of wave propagation (and therefore the speed of light). More precisely, the speed of waves in the elastic ether, given by the square root $\sqrt{A}$ of a constant term A, the ratio of "the elastic ether forces [...] to the ether masses to be moved by these forces", becomes $\sqrt{A + k}$ in the presence of a magnetic field. The k term, proportional to the additional elastic deformations, reflects the existence of the field. This is the only mathematical expression in the brief. Albert then points out a subtle point that requires knowledge of anisotropic media: if a wave arrives perpendicular to the field, then its speed must depend on the direction of its polarisation; then he feels it is necessary to specify that once the question of the effect of the "three components of elasticity" on propagation has been determined, it will be necessary to distinguish between cases where the field lines are parallel or emanate from a magnet pole.

The programmatic research that Albert mentions in his article is very probably linked to the question of the nature of electric current. He states that in

---

[116] Letter to Caesar Koch in *CPAE (English translation supplement)*, vol. 1, Doc. 6, pp. 6-7.

the case of the magnetic field created by a wire "the exploration of the elastic state of the ether in this case would permit us a look into the enigmatic nature of electric currents". It should be remembered that at that time there was no experimental proof that electricity in conductors was particulate in nature, even though in 1892 Hendrik Antoon Lorentz hypothesised that it consisted of charges free to move in the ether and in 1894 Joseph Larmor envisaged that the electron was the centre of a vortex that was conserved indefinitely. Other interpretations, such as the cracking of the ether, could also be envisaged[117]. The subject remained fresh in Albert's mind the following year, as his friend Ernestina Marangoni reported that in 1896 he had given her an article entitled "About electricity and electrical currents", which could be seen as a continuation of the dissertation. But he asked her to return it to him the following year "because it was wrong[118]", which suggests, on the one hand, that he may not have been referring to an electronic theory in this new writing (the electron was discovered at that time), and on the other hand, that it may have been intended for someone better placed to judge it than Ernestina, such as his uncle, the physicist Carlo Marangoni, to whom we shall return later.

If the aim of the 1895 dissertation was indeed to demonstrate a precocious aptitude for scientific reasoning, a knowledge of the properties of the elasticity of media and electromagnetism, a sense of problematisation and therefore a certain maturity, all in response to Herzog's request, perhaps it was achieved and Albert was given a chance to appear at an oral interview at the ETH, on the general knowledge section. Following his failure, he followed Herzog's initial advice, which was to complete his secondary education, by going to the cantonal school in Aarau to prepare for his Matura, which he obtained in 1896. In Aarau, he and his young cousin Robert Koch, his uncle's son living in Genoa, were taken in by Professor Jost Winteler, a friend of Gustav Maier, to whom Maier had written to ask him to admit the two cousins.

What are the sources for the dissertation?

At the time, electromagnetic waves were a fast-growing subject, occupying centre stage on the scientific scene. In 1892, Heinrich Hertz (1857-1894) published *Electric Waves; Being Researches on the Propagation of Electric Action with Finite Velocity through Space;* a collection of his publications on the discovery of electromagnetic waves since 1887, preceded by a long introduction that brought to life the different stages of his discoveries and his questions.

---

[117] Olivier Darrigol, *Electrodynamics from Ampère to Einstein*, Oxford, Oxford University Press, 2000.

[118] See Lucio Fregonese, *Gioventù felice in terra pavese*, *op. cit.*, p. 17, translation from Italian; and the conclusion of the article by Fabio Bevilacqua and Stefano Bordoni, "L'articolo del Giovane Einstein scritto nel 1895", in Fabio Bevilacqua and Jürgen Renn (Eds.), *Albert Einstein, ingegnere dell'universo*, *op. cit.*, pp. 276-279.

A reading note published in the *Elektrotechnische Zeitschrift* drew the reader's attention to this highly accessible and particularly interesting introduction. The year 1894 was also marked by Hertz's death on January 1, and several articles were devoted to him in tribute in electrical engineering journals. Oliver Lodge, for example, published five articles in *The Electrician* on Hertz's work. It was therefore not surprising to see the young Albert mention "the marvellous experiments of Hertz" at the beginning of his dissertation in 1895, which he had heard about if only through the magazines in his uncle Jakob's office.

In order to bring Albert's problem closer to a contemporary one, the *Collected Papers* mention Lodge's experiments designed to demonstrate a change in the speed of electromagnetic waves under the effect of ether entrainment or the presence of a magnetic field. The first results of Lodge's interferometric experiments[119], published in 1892 under the title "On the present state of our knowledge of the connection between ether and matter" in the *Proceedings of the Royal Society* in London, were reproduced in *The Electrician*. They created a controversy, as their conclusion was incompatible with the classical interpretation of Michelson's experiment in terms of total ether drag hypothesis and were debated by George Francis FitzGerald in the Royal Society, London, which was also echoed in the journal. Lodge then examined the additional effect of a magnetic field, either transverse or longitudinal, on the speed of propagation of light rays, but without further success. Joseph Larmor reported these negative results in 1893 at the end of a paper in which he developed a theoretical point of view on the dynamics of the ether and deduced as constraints "that motion in a magnetic field is very slow, and the density of the medium correspondingly great". His paper to the Royal Society was reproduced in *Nature* and presented in *La Lumière électrique* by Jules Blondin in May 1894.

Albert, for his part, worked on the elasticity of media with the physics courses given at the University of Lyon by Jules Violle, translated into German and published by Springer[120], purchased from the Hoepli bookshop in Milan. These books are still in his personal library, with numerous annotations, as the *Collected Papers* show. Maja Einstein recalls that her brother had worked on these volumes. Violle's name is also associated with electrotechnical circles: he was the editor of the report on the session devoted to electric lighting at the Paris Exhibition of 1881, as we saw in the second chapter, and had also published several articles in *La Lumière électrique* on the emission spectrum of incandescent lamps (blackbody radiation). Since the Einstein

---

[119] The beams of an interferometer passed between two turntables, separated by an inch, driven by a 9-horsepower Mather and Platt electric motor, which gave them a rotational speed of 4 000 rpm. This experiment is described as "heavy engineering" by Bruce Hunt, "Experimenting on the Ether: Oliver J. Lodge and the Great Whirling Machine", *Historical Studies in the Physical and Biological Sciences* **16** (1), 111-134 (1986).

[120] Jules Violle, *Lehrbuch der Physik*, Berlin, Springer, 1892 (vol. 1) and 1893 (vol. 2), German translation of the French works.

family had contacts with the academic world in Pavia, as we shall see, Albert could have turned to Prof. Carlo Somigliana (1860-1955), who was teaching in the third year, if he had needed further information on anisotropic elastic media, where Carlo was a renowned specialist.

A work that might also have caught his attention in electromagnetism was Paul Drude's *Electromagnetic Foundations of the Physics of the Ether.* It had just appeared in 1894, following the German translation of the first edition of Henri Poincaré's *Électricité et optique,* to which it refers. Drude's book is an introductory manual on magnets, currents, induction, etc., but it also gives a detailed account of Maxwell's electromagnetic theory. It should be noted that Drude describes the effects, well known to electromagnet users, of anisotropic magnetic field stresses on a medium and ends his analysis by suggesting a possible influence of the magnetic field on the ether (albeit a much smaller one). Albert, or someone in his scientific circle, may have been intrigued by this statement. Note that the letters to Mileva show that Drude's book was the first book Albert read in detail at ETH.

Perhaps Ernestina's uncle, Carlo Marangoni, who had not only worked on elasticity, but had also compared the propagation of light and that of an electric spark in birefringent media, under the title *New relationship between electricity and light,* could have given him some advice? If an explanation other than that of Albert's independent writing – based on scattered sources and various pieces of advice from engineers or academics in his environment – were to be considered, it could be that of an early meeting with Michele Besso in September 1895, for example, as we suggested in the previous chapter. Through his studies at ETH, which he had just completed, Besso knew both the authors mentioned above and the professors to whom the dissertation might have been addressed.

The Einsteins' links with the University of Pavia

Albert's family has links at the highest level with the University of Pavia. As a result, in his teenage years Albert was surrounded not only by engineers but also by academics, who could help him prepare for ETH. It should be noted that in his autobiographical sketch, in 1895, he wrote that his "incomplete knowledge" for his preparation for ETH came "mainly from personal study", implying that he also had other sources of training.

Their partner Lorenzo Garrone and the mathematician Giulio Vivanti

Hermann and Jakob Einstein were not the only investors in the Pavia company, and Lorenzo Garrone, whom they had known for several years, was their main partner, as we have seen. Born in Turin on January 28, 1858, he was the son of Giacomo Garrone and Margherita Arigo. He was living at via Carlo Felice 14 when he enrolled as a student at the Turin Polytechnic, after two years studying physics and mathematics at university. He graduated as a civil engineer in 1881, with the rank of seventieth out of eighty-five. At his school, he studied general chemistry, theoretical geodesy, graphic statics, descriptive geometry and rational mechanics in his first year; architecture, construction, technical physics, practical geometry, mechanics, hydraulics and legal subjects in his second year; plus courses on steam engines and railways, mineralogy and geology, mechanical technology and economics in his third year.

He belonged to the same class of engineers as the mathematician Giulio Vivanti (1859-1949), who was also in Pavia in June 1894, when he was appointed professor at the Normal School of the Faculty of Science. From February to April 1895, Vivanti replaced Professor Giacomo Platner in complementary algebra and analytic geometry for first-year students, and taught subjects that covered those of the ETH entrance exam syllabus in analysis and algebra. In 1910, he was appointed a member of the Lombard Institute, Milan's academy of science and letters, and in 1922 he was the rapporteur for Albert Einstein's appointment as a foreign member of this institute. Albert could certainly have followed his courses for ETH, in addition to what he saw for himself or with his uncle. Remember that a first year at university is a good preparation for entering the school, as Michele Besso had done in Rome under similar circumstances.

Considering that Albert was able to attend courses at the University of Pavia in the second half of 1895, the question that arises is whether Italian was an obstacle for him. Maja explained that her brother Albert had stayed in Munich while the family went to Milan, not only to complete his secondary education there, but also "because of the Italian language, which was foreign to the boy". When, a year later, he took the entrance exam for the cantonal school in Aarau in October 1895, he scored 3/6 in Italian, hardly different from 2.5 in German. However, he still had major gaps in his French, which he made up during his stay in Switzerland. An average mark indicates a decent level, because the requirements in Italian, as in French and German, are high at a Swiss school, as it is one of the country's official languages. His friend Ernestina Marangoni also said that Albert only had puns he needed to make in German. This implies that he had learnt the Italian language and its grammar as soon as he arrived in Milan, and that he practised it regularly during his stay in 1895.

Their partner Angelo Cerri, theoretical geodesy assistant

When the Einsteins set up their company in Pavia, columnist Fabrizio Bernini wrote that the engineer Angelo Cerri (1848-1902), their second partner, was "a university assistant and lecturer (*docente*) at the Pavia

Technical Institute[121]". We also learn that he was related to the family of Davide Giulietti, the lawyer who set up the company, and therefore belonged to the Venco family, the maternal branch of Ernestina Marangoni's family.

Angelo, son of Siro Cerri, was born in Dorno. The historical archives of the University of Pavia hold his student and teaching records. He began his higher education at the University of Turin in 1867, where he obtained excellent marks in complementary algebra, analytical geometry and chemistry. Experimental physics was taught by Gilberto Govi, one of Italy's representatives at the 1881 congress of electricians in Paris. Cerri then enrolled for second- and third-year mathematics courses in Pavia, and in 1869 attended Giovanni Cantoni's experimental physics course. He brilliantly obtained his bachelor's degree in pure mathematics in 1870, with 30/30 in theoretical geodesy, 28/30 in differential and integral calculus and descriptive geometry and 27/30 in physics and rational mechanics. From 1885 to 1900, he held a yearly renewable position at the University of Pavia as first assistant in geodesy under Prof. Girolamo Gobbi Belcredi, whom he replaced on his death in 1898. He was then appointed professor of topography at the Royal Technical Institute in Chieti and applied unsuccessfully to the Polytechnic in Milan.

The archives of the Antonio Bordoni Technical Institute in Pavia show that Cerri also taught French at the business school's evening classes from 1886 to 1888. During 1895-1896, he also helped with topography exercises. Between 1894 and 1896, he published three scientific articles in the journal *Il Politecnico* and an article in the proceedings of the Lombard Institute on the multiple reflections of light rays in optical devices. His students Lodovico Aceti and Moisè Tedeschi published his *Lectures on Theoretical Geodesics* in 1900. The lectures covered spherical trigonometry, geometrical optics and its application to the theodolite, the correction of atmospheric refraction, the geometry of surfaces in three-dimensional space and the notion of the geodesic line. This link to geodesics was crucial in an engineering community involved in the development of railways. Perhaps this early relationship, in this milieu of engineers and academics, predisposed Albert to the use of a more sophisticated geometry for his analysis of gravitation in 1915, in which the notion of the geodesic line played a central role. Albert could in any case have been discussing trigonometry and optics with Cerri as early as 1895, with a view to the ETH.

---

[121] Fabrizio Bernini, *Che bel ricordo Casteggio... : Albert e Maja Einstein ed il salotto letterario di Ernestina Pelizza Marangoni*, Edizioni Modulo Tre di Diego Dabusti & C., 1994.

Jakob Einstein, Otto Neustätter and the medical academics in Munich and Pavia

Jakob Einstein had links at the highest level with the University of Pavia, even before the company moved there. In the anniversary letter to Albert mentioned above, Otto Neustätter began by thanking Jakob for his study stay in Pavia: "Your uncle, to whom I owe the interesting apprenticeship in Italy [...]". This means that Jakob was probably involved in obtaining his assistant's post in Pavia. During the academic year 1894-1895, Otto Neustätter was an assistant to the professor of ophthalmology Francesco Falchi (1848-1946). On December 29, 1894, Falchi had written to the Rector of the University of Pavia, the future Nobel Prize winner for medicine Camillo Golgi, asking him to take on Dr Neustätter as a second assistant in his clinic. Neustätter was appointed to the ophthalmology clinic by decree on January 4, 1895, with a contract worth 1,000 lire, renewable annually until October 31, 1897. He subsequently wrote to Golgi asking for his salary to be paid retroactively from the first lessons he had given in November. This makes it possible to date his arrival in Pavia to autumn 1894, shortly after the Einsteins moved there in the summer. A year later, in November 1895, Neustätter wrote to Professor Falchi to tell him that he was resigning because his father had given him a job as an assistant at the University of Königsberg, and that he would not be able to return to Italy for some time. The Rector of the University of Pavia acknowledged his resignation by decree on November 21, 1895[122].

Otto Neustätter graduated in medicine from the University of Munich in 1894, so his connection with Jakob Einstein goes back to the family's Munich days. It involves Jakob's relationship with the hospital environment. Jakob had been asked by Oskar von Miller to take charge of part of the International Electrical Exhibition in Munich in 1882 and must therefore have been in contact with the director of the Munich hospital, the electrophysiologist Hugo von Ziemssen, who, like Miller, had been one of the fifteen German representatives at the 1881 congress of electricians in Paris. For his part, Neustätter was to follow the courses of Ziemssen, who published his *Clinical Lectures* in twenty-seven volumes between 1887 and 1900. Ziemssen was probably the link between the Einsteins, Neustätter and Pavia's renowned medical community.

In Pavia, in October 1895, the Einsteins won the contract to light the Palazzo Botta, which houses the Faculty of Medicine. The contract stipulated

---

[122] Neustätter later organised the History of Hygiene Exhibition in Dresden in 1911. In 1939, he wrote again to Einstein on his sixtieth birthday. He used a letterhead from the Institute for the History of Medicine at John Hopkins University in the United States, where he had stayed from 1937 to 1939 after fleeing the Nazi regime, *Albert Einstein Archives* online, doc. no. 30-631.

that the company would provide 120 lamps for the human anatomy, comparative anatomy, hygiene, forensic medicine and general pathology laboratories and that it would supply part of the electricity needed for the lighting. It should be noted that on January 16, 1896, Camillo Golgi informed the company that breaches of the contract had been noted, for example that incandescent lamps had been replaced by arc lamps without justification, which he diplomatically attributed to a possible translation error in the documents. The text of the contract and the letter are published in Lucio Fregonese's book. It should be noted in passing that the legal entity representing the University at the signing of the contract was the engineer Eugenio Franchi Maggi, married to Bianca Casorati, the second daughter of the mathematician Felice Casorati (1835-1890), who taught in Pavia and at Milan's Polytechnic, once again highlighting the intertwined links between the networks of engineers and academics.

Links with physicists at the University?

Science historian Fabio Bevilacqua noted that the Einstein family lived in Pavia just a hundred metres from the house of Adolfo Bartoli (1851-1896), the professor of experimental physics and director of the university's physics laboratory, a corresponding member of the Lombard Institute. Bartoli's name has remained associated with the pressure exerted by light on a metal in Maxwell's electromagnetic description (Maxwell-Bartoli pressure), a subject that would recur several times in Albert's work between 1905 and 1916, in connection with his central theme of statistical fluctuations. More generally, Bartoli was a specialist in thermal measurements and occasionally worked for major companies in the electrical sector[123].

The possible link with physicists leads us to mention Giovanni Cantoni (1818-1897), a major player in physics in Pavia, who was succeeded by Bartoli in 1893. Cantoni played an important role in the field of electricity, which could provide a link with the Einstein enterprise. Let us start with a few biographical details. Giovanni Cantoni took part in the *Risorgimento* and the Five Days' Uprising in Milan. When the Austrians returned to the Lombard capital, he went into exile and joined Carlo Cattaneo in Lugano, at the technical college the latter had founded, as we saw in the first chapter. There he wrote *Elements of Physics for Secondary School Use.* Appointed professor at the University of Pavia in 1860, he became Rector between 1862 and 1868. He was also a member of the Lombard Institute from 1862 and was elected Senator of the Kingdom of Italy in 1879. In the field of electricity, he was one of the leading contributors to the Florentine journal *Elettricista* and

---

[123] Emilio Gatti and Alberto Gigli Berzolari, "Fisica", in *L'Istituto lombardo, Accademia di scienze e lettere*, vol. 2, *Storia della classe di scienze matematiche e naturali*, pp. 146-147.

was a member of the preparatory committee for the Paris Exhibition of 1881, for which he and Rossetti wrote the Italian Bibliography of Electricity. In his tribute to Cantoni at the Lombard Institute in 1899, the physicist Oreste Murani, of the Milan Polytechnic, listed around two hundred publications[124]. His scientific work focused on electromagnetic induction, elasticity, molecular physics, osmotic pressure, the diffusion of liquids through porous membranes as a function of temperature and Brownian motion[125].

Did Albert's interest in the kinetic interpretation of Brownian motion arise in Pavia between 1895 and 1900? Albert specified in his interviews with the physicist Robert Shankland that the explanation of the phenomenon in 1905 came to him suddenly, but this does not contradict an earlier knowledge of this interpretation. In the founding issue of the history of science journal *Isis* in 1913, Icilio Guareschi gave an account of the phenomenon discovered by the botanist Robert Brown in 1827 concerning the random agitation revealed under the microscope when observing pollen grains in solution; he wrote: "As early as 1867, Professor Giovanni Cantoni had clearly recognised that Brownian motion depends on the movement of the molecules of the liquid in which the substance is in suspension[126]". In his 1867 article, Giovanni Cantoni used Dulong and Petit's and Graham's laws of specific heat and diffusion to deduce that the mean molecular kinetic energy $mv^2/2$ is a constant that depends only on temperature; he then applied this result to Brownian particles in liquids and concluded that their motion is "one of the most beautiful and direct experimental demonstrations of the fundamental principles of the mechanical theory of heat[127]". It is highly likely that Albert became aware of this vision of Brownian motion very early on through discussions, not with Cantoni directly, but rather with Carlo Marangoni, his former student and assistant[128]. Albert's merit in May 1905 was that he went further than Cantoni in analysing the process and explicitly calculated the mean free path of particles as a function of their size from their diffusion coefficient: the application of a general result that he had just obtained in his thesis defended at the end of April.

---

[124] Oreste Murani, "Commemorazione del sen. prof. Giovanni Cantoni", *Rendiconti Reale Istituto Lombardo di scienze e lettere* **32**, 57-107 (1899).

[125] Emilio Gatti and Alberto Gigli Berzolari, "Fisica", art. cit.

[126] Icilio Guareschi, "Nota sulla storia del movimento browniano", *Isis* **1**, 47-52 (1913).

[127] Giovanni Cantoni, "Su alcune condizioni fisiche dell'affinità, e sul moto browniano", *Il Nuovo Cimento* **27**, 156-167 (1867); and *Rendiconti Reale Istituto Lombardo* **1**, 56-67 (1868).

[128] The journal *Il nuovo Cimento* was at that time directed by the professors Matteucci and Piria, but also "continued by the professors of physical and natural sciences of Pisa and the Museum of Florence" as detailed on the front page of the volume. Among them, there must have been Carlo Marangoni, since he was Matteucci's assistant at the Florence Museum.

Social relations in Pavia: Ernestina Marangoni and her uncle Carlo, a physicist

Ernestina Marangoni was a mutual friend of Albert and his sister Maja in Pavia, with whom she kept up a lasting correspondence. She would see her regularly, since Maja moved to the Florence region in 1922 with her husband, Paul Winteler, before joining Albert in the United States after the anti-Semitic law passed in Italy in 1938. In the 1910s, Ernestina ran a literary salon in Casteggio that was frequented by her first cousin, the art critic Matteo Marangoni (1876-1958), son of Carlo Marangoni. The Italian poet Eugenio Montale (1896-1981), who won the Nobel Prize for Literature in 1975, was also a close friend of the Marangonis, having lived with them in Florence and accompanied them to their villa in Casteggio in the autumn. The link with Ernestina testifies to the good social relations that the Einstein family maintained with the bourgeois and intellectual milieu of Pavia. But it also opens up a hitherto unknown aspect of Albert's relationship with Ernestina's uncle, the physicist Carlo Marangoni, which may have had a direct impact on his early scientific interests.

Ernestina Marangoni

It was through Otto Neustätter that Albert met Ernestina Marangoni (1876-1972). Fabrizio Bernini, a historical chronicler of the Oltrepo region, got to know Ernestina when she was approaching her ninetieth birthday (he was sixteen at the time) and published the booklet *What a beautiful memory, Casteggio...*, which provides interesting details of Albert's ties to the family. He refers to a note from Ernestina in 1949, which recounts the circumstances of their first meeting:

> It was 1896, and my mother and I were in Pavia, bathing in the Ticino. That's when a family friend came to visit us, Otto Neustätter from Königsberg, a professor who had come to practise ophthalmology with the illustrious Professor Falchi, and who was accompanied by a young man whom my mother was introduced to as Albert Einstein. At that very moment I was learning to 'wakeboard' in the waters of the Ticino [...]. I made my way to the shore where I had the opportunity to make the acquaintance of this young Einstein, who was seventeen years old.

In 1955, however, Ernestina wrote in the journal *Provincia Pavese* that the meeting had taken place in the summer of 1895 (and not 1896), which seems to be borne out by the fact that Neustätter left Pavia at the start of the academic year in 1895, as we saw earlier.

Ernestina was born in Casteggio, a small town 25 km south-east of Pavia, in November 1876. Her father, Giulio, born in Pavia in 1843, was an expert in raising silkworms used in the textile industry; he died in 1906. Albert clearly held him in high esteem, comparing him to a "Leonardo da Vinci" in a letter to Ernestina. Her mother, Rachel Venco, came from a family that had specialised in pharmacy for several generations. In Pavia, Albert accompanied Ernestina on the violin as she played the piano, and the friends met up every week. After entering the Bordoni Institute in 1895 – where Cerri taught – she obtained her doctorate in pharmacy from the University of Pavia in 1902; her future husband, Edmondo Pelizza, had obtained his PhD the previous year. The family took part in the *Risorgimento*, and in 1934 Ernestina published a book in tribute to this period, entitled *A small Garibaldian world, Alba Coralli Camozzi, her family – her friends.*

The Pavia University Museum holds three handwritten letters from Albert Einstein to Ernestina Pelizza Marangoni, dated August 16, 1946, October 7, 1947 and October 1, 1952. They are published by Lucio Fregonese, with their translations, preceded by an introduction and followed by the reproduction of an article by Elena Sanesi on the establishment of the Einsteins in Italy. After the Second World War, Ernestina contacted Albert again to ask for his support in rebuilding the covered bridge over the Ticino to the south of Pavia, which had been destroyed during the war. Today, a commemorative plaque marks Albert Einstein's passage over the bridge. These letters bear witness to the friendship and fond memories Albert retained from his youth of living with the Marangoni family in Casteggio. In 1946, he wrote to Ernestina: "The happy months of my stay in Italy are my fondest memories", and in 1952: "What a beautiful memory, Casteggio", the title of Bernini's book. When he visited them in Casteggio, he stayed in an outbuilding of their magnificent and immense villa, built high up on the *Pistornile* at the beginning of the 19th Century in the style of a Venetian palace. Albert stayed there in particular during the traditional grape harvest festival in September, from 1895 to 1900, during which he had made a name for himself in the surrounding countryside, where he was nicknamed "the German".

Carlo Marangoni, a renowned physicist and teacher

Carlo Marangoni (1840-1925) was Ernestina's uncle. Although he had been living in Florence for several years by 1895, he visited his family in Casteggio once a year, like Albert, during the grape harvest, according to a private communication from Fabrizio Bernini. He did not stay at the Villa Marangoni, as he wanted more peace and quiet at this festive time of year but stayed at a nearby farmhouse in the middle of the vineyards, on an estate belonging to the Pelizza family. Carlo was bound to make friends with a young man who wanted to study physics at ETH, especially as in 1879 he himself had written a textbook on experimental physics for future teachers[129]. We will

---

[129] Carlo Marangoni's best-known books, listed in his obituary in the journal *Il Nuovo Cimento*, include *Liquid Bubble Monograph* (1873), which he published with Pietro Stefanelli, his physics textbook *Experiments and observations as a complement*

take a closer look at his education and his research interests before turning to possible topics for discussion with Albert.

Carlo Marangoni was the older brother of Giulio, Ernestina's father, as confirmed by the diocese of Pavia in the parish registers of San Michele for the years 1840 and 1843: Carlo (Luigi Giuseppe) and Giulio (Giovanni Battista) were the children of Matteo Marangoni, whom Carlo described as an accountant in his enrolment form for the University of Pavia, and Ernesta Robecchi. Carlo completed his third year of studies in pure mathematics by obtaining a doctorate in mathematics from the Faculty of Physical, Mathematical and Natural Sciences on December 18, 1863. His diploma bore the signature of Giovanni Cantoni, then Rector of the university. His inaugural dissertation was entitled "On the rise of sap in plants", in which he compared the theories of Jamin and Dutrochet on the rise of sap in relation to the phenomena of endosmosis.

It was with Cantoni that he wrote his thesis entitled *On the expansion of a drop of liquid on the surface of another liquid*, which he defended in 1865 and which he places under the sign of the famous French saying: "Ne cherchez pas midi à quatorze heures" [Do not seek noon at 2 pm; *i.e.*, don't indulge in nitpicking]. By meticulously observing the successive stages in the expansion of a drop of essential oil on water, he came to refute previous interpretations that called on gravitational effects, the expansion of a gas, an electrical action or even an unknown force to explain the phenomenon. He established that the expansion of the drop was solely linked to the surface tension of the liquid, the same as that responsible for the phenomena of capillarity, which can be observed when a liquid rises along the wall of a very thin glass tube, for example. He observed that this tension was modified by the presence of a drop of a different kind of liquid and gave a detailed account of his experiments in the great pond at the Tuileries Gardens in Paris.

In 1865, Carlo became Cantoni's assistant, whose lectures he published in his *Summary of physics lessons for the year 1865-66*, and then that of Carlo Matteucci (1811-1868) at the Florence Museum of Physics, which, along with medicine and philology, was one of the three components of the city's Technical Institute. He then became professor of physics at the Lyceum Dante in Florence from 1870 until his retirement in 1916, but continued to work at the Institute, where he pursued his research, as can be seen from the dedication to Professor Antonio Roiti of his article "Paramagnetism and diamagnetism" (1885), which mentions the Institute. His other scientific works include those on "surface viscosity", introduced on the basis of capillary phenomena[130], the plasticity of solids, crystallography, the propagation

---

to physics textbooks (1879), intended for future physics teachers, and *On the swim bladder of fish* (1880) (titles translated from Italian).

[130] He was involved in a controversy on this subject for several years with the Belgian physicist Joseph Plateau.

of an electric spark in minerals, hail, the principle of several physics' instruments and the new relationship between electricity and light mentioned in connection with Albert's 1895 article[131].

Carlo Marangoni: a direct influence on the young Albert?

Carlo Marangoni dealt with capillarity in his 1879 educational work. In particular, he drew up a table of molecular distances that could be deduced from this phenomenon by various authors and attributed the most accurate measurement 1/250 000 mm to Jules Violle. His name in physics is still attached to the Gibbs-Marangoni effect, which, following on from his thesis, describes the appearance of a current of matter when two liquids with different surface tensions come into contact. It's tempting to imagine him explaining the phenomenon of 'wine tears', for example, to a young budding physicist, on the basis of his work, during a toast given at the grape harvest festival! The phenomenon is explained by the fact that, close to the wall of the glass, the wine rises by capillary action: in this zone, the evaporation of the alcohol causes a gradient in the surface tension of the liquid (that of the alcohol being lower than that of water), causing a displacement of matter towards the zones of greater tension; the wine rises along the walls then sinks back down under the effect of its weight, hence the phenomenon observed.

It should be noted that in his 1907 encyclopaedic article on capillarity, Hermann Minkowski quotes Marangoni three times[132], which shows that he was indeed a recognised specialist in the subject. Since 1897, Minkowski had been giving a course on capillarity at ETH, which Albert, according to his classmate Louis Kollros, appreciated because it was their first course in mathematical physics. It may also have given him the opportunity to learn more about a subject with which Carlo had familiarised him. It was on the basis of the use of capillary phenomena to determine a characteristic value for the attraction between atoms that Albert submitted his first article to the *Annalen der Physik* in December 1900, which was published in March 1901 and to which we shall return in Chapter 6. It should be noted that he cited Roberto Schiff for his experimental data relating to Ostwald's book. Schiff is a specialist in capillary phenomena. He was also a colleague of Davide Besso, Michele's uncle, at the University of Modena. His father, Moritz Schiff, is a thyroid specialist; he was professor of physiology and

---

[131] For a full account, see Emilio Borghi, "L'attività scientifica e didattica di Carlo Marangoni nel R. Liceo Dante di Firenze", *Atti della Fondazione Giorgio Ronchi* **3**, 287-332 (2008).

[132] Hermann Minkowski, *"Kapillarität"*, in *Enzyklopädie der mathematischen Wissenschaften mit Einschluss ihrer Anwendungen*, Bd. 5, *Physik*, edited by Arnold Sommerfeld, Leipzig, Teubner, 1903-1926, Teil 1, p. 558-613.

geology at the Florence Technical Institute, where he had also been invited by Matteucci, and was therefore a colleague of Carlo's.

Finally, it should be noted that the discussion imagined between Albert and Carlo on wine tears may also have sparked Albert's interest in the *Popular Lectures* by the famous British physicist William Thomson (Lord Kelvin), works that can be found in his library in their original edition[133]. The first volume of the collection (of which there were three) was published in 1889 and entitled *Constitution of Matter*. It begins with a seventy-three-page chapter on capillary forces. Thomson points out that it was his brother James who was the first to give an explanation of the phenomenon of the expansion of a liquid drop in 1855 and of wine tears[134]. It would be surprising if Carlo, who was attached to educational works, had not encouraged Albert to read Thomson and at the same time given him his point of view on this dispute over priority. Let us add a remark by Ernestina, which seems to lend credence to the idea that Albert may have read this collection of lectures before 1900. In her notes, quoted by Bernini, she recalls the sometimes-pensive attitude Albert adopted in society and links it to a particular memory: "Sometimes he would plunge into some meditation, as when smoking a cigarette, he would observe the rings of smoke and wonder about the physical laws that could make the phenomenon possible". In his book, Thomson extensively discusses smoke rings and, more generally, vortices in fluids, subjects initially dealt with by Helmholtz. Their properties are described in detail and he even envisages that vortices in the ether that last indefinitely could model atoms. Thomson recounted the experiments of his colleague Peter Guthrie Tait, who had created a device to illustrate the formation and behaviour of smoke rings using electric light, in theatrical presentations that were a great success:

> Lastly, large smoke rings discharged from a circular or elliptic aperture in a box were rendered visible, by aid of the electric light, in their progress through the air of the theatre. Each ring was circular, and its motion was steady when the aperture from which it proceeded was circular, and when it was not disturbed by another ring. When one ring was sent obliquely after another the collision or approach to collision sent the two away in greatly changed directions, and each vibrating seemingly like an India-rubber band. When the aperture was elliptic each undisturbed ring was seen to be in a state of regular vibration from the beginning, and to continue so throughout its course across the lecture-room. Here, then, in water and air was elasticity as of an elastic solid, developed by

---

[133] Sir William Thomson, *Popular Lectures and Addresses*, vol. 1, London, MacMillan & Co, 1889.

[134] In the appendix to this first chapter, Thomson even reproduces his brother James's paper to the British Association in Glasgow entitled "On certain curious phenomena observable on the surface of wine and other alcoholic liquors".

mere motion. May not the elasticity of every ultimate atom of matter be thus explained? But this kinetic theory of matter is a dream, and can be nothing else, until it can explain chemical affinity, electricity, magnetism, gravitation, and the inertia of masses (that is, crowds) of vortices.

If Albert had read Thomson's text around 1896, one wonders whether the article addressed – and then withdrawn – to Ernestina on the nature of electric current might not have had something to do with vortices in the ether, as Larmor did in 1894.

5
Albert's scientific environment in Milan

In October 1896, after spending a year at the cantonal school in Aarau to obtain his Matura, Albert finally was admitted to ETH in Zurich, without having to sit an exam, as his student handbook states. The canton of Aargau, where he studied, depended on the canton of Zurich for the pursuit of university studies and in 1895 Albert had been assured of admission to the ETH the following year, according to his sister Maja's own recollection. In Milan, his family moved permanently to 21 via Bigli, after the sale of the Einstein, Garrone & Cie. On February 23, 1899, Hermann, his father, set up a new company under the name Einstein & Cie, for which he was the sole signatory. Registered with the Milan Chamber of Commerce on July 4, 1899, it was presented as the result, for reasons of "greater opportunity", of the transfer of the Pavia company founded in 1894. The workshops were located at 160 via Lecchi and the administrative office was registered at the family home.

During his time at ETH, Albert stayed regularly and for long periods at his parents' house, during the school's half-term breaks in spring (March-April) and summer (August-September), as well as at Christmas. Traces of his stays can be found in the letters he wrote to his girlfriend Mileva Marić, whom he married in January 1903 and to whom he wrote regularly from Milan. These letters allow us to follow the development of his scientific questionings from 1898 to 1901, between his nineteenth and twenty-first birthdays, thanks to the many references that punctuate them. This was an important period for the young Albert, who was first preparing his diploma (*Diplomarbeit*) for ETH and then working on his first thesis at the end of 1900 and during 1901. According to Michele Besso, he was a brilliant and passionate young student who was fascinated by the essential questions of physics, such as the existence of the ether and atoms. Like any student in his position, he carried out in-depth bibliographical work on his research topics.

During his stays in Milan, Albert devoted his free time to studying. In September 1900, he wrote to Mileva: "My only distraction is studying". He opened up about his scientific questions in almost daily discussions with Michele, which led him to write: "I am spending many evenings here at Michele's". As we shall see, the letters to Mileva enable us to identify the library in which Albert carried out his research with that of the Lombard Institute, the Academy of Science and Letters, an institute we have already encountered on several occasions. They also show that he was working for Michele on wireless telegraphy, a highly topical subject at the time. He even encouraged his friend to undertake a thesis, which we will link to this theme through our discussion of the contents of the library of Giuseppe Jung, Michele's uncle. To illustrate Albert's approach to scientific questions, we will consider the case of the specific heat of bodies (variation of their internal energy with temperature) and their thermal conductivity (transport of heat). We will then take a broader look at his scientific and social environment, when he shared his family's entrepreneurial concerns, while applying for numerous university assistantships. We shall see that in Milan he benefited from protections in the scientific field, with Giuseppe Jung and, probably, Luigi Gabba in connection with Bernardo Ansbacher, a friend of his father's whom he met again in his musical environment.

The library of the Lombard Institute (1899-1901)

In September 1899, Albert wrote to Mileva from Milan: "I too have done much bookworming & puzzling out, which was in part very interesting[135]". We can therefore assume that he was not only working on his own works or on those sent to him by Mileva from Zurich, but also that he was conducting his research in Milan in a scientific library. This point was confirmed in 1901 by the concluding sentence of his letter dated April 4: "But I must be off to the library, otherwise it will be getting there too late[136]". Albert had therefore become accustomed to working in a Milanese library. But which one? Let's start by gathering more details about his reading from his correspondence with Mileva, which will direct us to the various titles related to the *Annalen der Physik*.

After writing in 1899 that doing "...much bookworming & puzzling out ..." he had found some "very interesting" things, he went on to say: "I also wrote to Professor Wien in Aachen about the paper on the relative motion of the luminiferous ether against ponderable matter, which the "Principal" [Weber] treated in such a stepmotherly fashion. I read a very interesting paper published by this man on the same topic in 1898". This is Wien's article, entitled "On questions relating to the translational motion of the luminiferous ether", published in volume 65 of the *Annalen der Physik und Chemie* in 1898. It appears as an appendix (specifically numbered in Roman numerals) and is therefore not listed in the journal's index. This characteristic makes it particularly difficult to find, as it can be placed at the beginning, end or between two consecutive monthly issues of the annual volume, its position depending

---

[135] Letter to Mileva Marić in *CPAE (English translation supplement)*, vol. 1, Doc. 57, p. 135.
[136] Letter to Mileva Marić in *CPAE (English translation supplement)*, vol. 1, Doc. 96, pp. 162-163.

on the choice of the binder. This in itself requires a bit of research[137]! We can assume that Albert had just read this article from Wien and therefore had access to the *Annalen der Physik und Chemie* in Milan.

In his letter of April 4, 1901, Albert also stated that he "had in [his] hands a study by Paul Drude on the electron theory" and confirmed on April 10 that "Last week I studied electrochemistry and chemical reactions from Michele's 'Ostwald', and the electron theory of metals in the library[138]". According to the *Collected Papers*, these are two articles published in Drude's monthly journal *Annalen der Physik*, which took over from *Annalen der Physik und Chemie* in 1900.

The letters also contain other indications about the journals he read, such as when, again on April 10, 1901, Albert wrote, after expressing his doubts about Planck's vision: "Maybe his newest theory is more general. I intend to have a go at it". Planck's talk to the German Physical Society on December 15, 1900, in which he introduced the quantification of radiation energy on material oscillators, was summarised in the March 1901 issue of *Beiblätter zu den Annalen der Physik*. It was also published in the March issue of *Annalen der Physik* at the same time as Albert's first article on molecular forces. We can therefore deduce that either the issue of the journal in question is in the library and Albert has only skimmed through it, or that he has read the announcement of the paper delivered in December and the article to appear in the *Beiblätter*.

To sum up, we are looking for a library in Milan that holds the three titles related to the *Annalen*: *Annalen der Physik und Chemie*, *Annalen der Physik* and *Beiblätter zu den Annalen der Physik*. Only the library of the Lombard Institute, the Academy of Sciences and Letters, had these journals in Milan at the time[139]. We will see in the next chapter that a fourth important source appears as a watermark in the letters to Mileva sent from Milan, the *Jubilee* for Lorentz, which is published in volume V of the *Archives Néerlandaises des Sciences Exactes et Naturelles* (*Dutch Archives of Exact and Natural Sciences*). It can also be found in this library. Albert had several options for gaining access to this academic library. He could, for example, involve Giuseppe Jung, Michele's uncle and a member of the Institute, whom he told Mileva in March 1901 would provide him with connections in Italy, or even ask Giuseppe Colombo directly, who had been president of

---

[137] This text from Wien reproduces a paper given at the seventieth Conference of German scientists and physicians in Düsseldorf. In it, he describes Lorentz's electrodynamic theory of 1895, to which we will return in the next chapter, and describes thirteen experiments that can be explained by this theory, which will have been of great interest to Albert.

[138] Letter to Mileva Marić in *CPAE (English translation supplement)*, vol. 1, Doc. 97, p. 163.

[139] The physics library at the University of Milan only holds the *Annalen der Physik* from 1900 onwards, but these are reproductions acquired at a later date, as the university was not founded until 1924.

the Lombard Institute and knew his parents' business well, as we saw in the first chapter. Remember that to access the library of the Institut de France today, the model on which the Lombard Institute was founded, you need the recommendation of an academician.

While the library of the Lombard Institute is very rich in periodicals, it is less so in university textbooks, particularly in German. For example, the book by Ostwald mentioned above does not appear in the library (which may explain why Albert borrowed it from Michele), nor do works by Helmholtz, Boltzmann or Mach, which he asks Mileva to send him from Zurich to Milan in September 1899. There is therefore a strong correlation between the Institute's library holdings and all the bibliographical items appearing in the letters to Mileva. Of course, the journals we are looking for are also to be found at the University of Pavia, the historic university centre of Lombardy, where Albert lived and where he could still consult them during his occasional visits. But the end of his letter to Mileva, in which he tells her that he must hurry to get to the library before it closes, shows us that it is in Milan, near his home, that he works. He could also have gone to the library of the Milan Polytechnic. Perhaps he also frequented it, but it did not hold the journals he was interested in, since the school had been set up at the instigation of the Lombard Institute and the two institutions were only a few hundred metres apart, so there was no reason for the journals to be duplicated.

It is not surprising to find the major scientific journals of the time in this library, among a collection that today numbers some 450 000 volumes and periodicals. A brief history of the Institute, created by Napoleon Bonaparte and whose first president was Alessandro Volta, was referred to in the first chapter. We have seen on several occasions that it played a key role in the development of Italian industry and research at the end of the 19th Century. Together with the Institute, the library was housed in the Palazzo Brera, a landmark of Milanese culture, five hundred metres from the Einsteins' residence at 21 via Bigli. Rudolf Kayser, Albert's son-in-law, bears witness to the fact that Albert frequented these premises in his 1931 biography:

> Milan was a paradise of freedom and beauty. He read much now, with that complete passion and devotion with which young men read [...] For the first time in his life, he studied the plastic and graphic arts: the last supper of Leonardo da Vinci at Santa Maria delle Grazie, the collections at Brera – a world of classical beauty[140]!

Until now, this statement has reflected a cultural and tourist image. For us, it confirms that Albert worked in the Institute's library in Brera, sometimes taking the time to admire the collections in the Pinacoteca on his way out of the library!

---

[140] Anton Reiser (pseudonym of Rudolph Kayser), *Albert Einstein: A Biographical Portrait*, London, Thornton Butterworth, 1931, p. 42.

Michele's work on wireless telegraphy

To discuss Albert's work for Michele, we are basing ourselves on a letter addressed to Mileva at the end of August/beginning of September 1900 from Milan, on which he adds, in the margin of the third page:

> I am investigating the following interesting problem for Michele: How does the radiation of electric energy into space take place in the case of a sinusoidal alternate current? About the amplitude of the waves produced as a function of the frequency of vibration, etc.[141].

The subject is wireless telegraphy, *i.e.*, the generation and propagation of electromagnetic waves. Let's start by recalling the background to the recent discovery of electromagnetic waves, of which we saw in the previous chapter that Albert already had some knowledge in 1895.

The discovery of electromagnetic waves

The existence of electromagnetic waves was predicted by the Scottish scientist James Clerk Maxwell (1831-1879) in 1866, based on his theory of electrical and magnetic phenomena. To describe the interaction between two electric charges or between two magnets, Maxwell, following in the footsteps of Michael Faraday, used the idea of a medium, the electric and magnetic ether, which would propagate the interactions progressively and which he illustrated with a mechanical analogy. The speed of propagation of disturbances in the ether, calculated from electrical and magnetic constants measured in the laboratory, is very close to the speed of light. What's more, the medium imagined by Maxwell can propagate waves the elongation of which is transverse to their direction of propagation, like light. For Maxwell, the luminiferous optical ether and the electric and magnetic ether are one and the same and light is an electromagnetic wave, described by very high-frequency oscillations of fields that are self-sustaining by induction. However, direct detection of electromagnetic waves had to wait another twenty years, with the experiments carried out from 1887 by Heinrich Hertz. The incident wave is generated by a high-frequency exciter (an oscillating circuit producing sparks in the air) consisting of a generator, a coil and a capacitor. The electromagnetic waves produced reflect off a metal plate fixed to the wall of the room where the experiments are being carried out, superimposing themselves on the incident wave to form a standing wave. The distance between two consecutive nodes of this wave is equal to half a wavelength, which is the product of the period of oscillation and the speed of propagation. The presence of nodes alone establishes the existence of electromagnetic waves

---

[141] Letter to Mileva Marić in *CPAE (English translation supplement)*, vol. 1, Doc. 74, pp. 147-148.

and invalidates the former electrodynamic theories, which were based on the idea of instantaneous action at a distance.

However, Hertz's experiments were unsatisfactory on one essential point. Maxwell's theory states that the speed of the waves generated in the conducting wire and in the air must be equal to the speed of light. But Hertz observed different values for the waves in the air and those in the wire. Henri Poincaré expressed his dissatisfaction in the second volume of the first edition of *Electricity and Optics,* published in 1891, which transcribed the lectures he had given in 1890 at the Sorbonne and which had a certain echo across the Rhine. Translated into German by Helmholtz's assistants in Berlin, it received a very positive review in the *Elektrotechnische Zeitschrift*[142] in 1892. While emphasising the universal scope of Hertz's work, Poincaré corrected him for a factor $\sqrt{2}$ in the measurement of wavelength in air due to Hertz's incorrect determination of electrical capacitance and pointed to several other possible causes of error. Hertz mentioned these remarks in the introduction to his book and Poincaré returned to them in the lectures he gave in 1892-1893, published in 1894 under the title *Electric Oscillations,* one chapter of which was devoted to Hertzian oscillations.

From his point of view, it was only the "new experiments by Sarasin and de la Rive", the Genevan physicists with whom Poincaré was in correspondence[143], that made it possible to affirm that "the equality of propagation speeds in wires and in air has been established". They carried out their experiments using the same protocol as Hertz, but in a room free of parasitic reflections, with a larger metal reflector that gave them more reliable measurements at long wavelengths. Their report to the Paris Academy of Sciences was published in *The Electrician* in 1893. These results marked the beginning of wireless telegraphy. The pioneers in this field were the Englishman Oliver Lodge and the Italians Augusto Righi and the self-taught inventor Guglielmo Marconi, winner of the Nobel Prize for Physics in 1909, who had studied with Righi at the University of Bologna. Experimental research focused on the creation of oscillators designed to generate high-frequency waves: Righi exciters, Lodge exciters, Bose exciters and so on. From a theoretical point of view, the aim was to determine the amplitude of the wave radiated with the frequency according to the devices used, which was precisely the work that Albert carried out for Michele in September 1900. It should also be noted that when Albert had contacted Giuseppe Jung, through Michele, in his search for an assistant's post, he reported to Mileva on April 4, 1901: "He promised that he will write to the most important professors of Italy (physicists), Righi and Battelli, on my behalf, *i.e.*, ask them whether they need an assistant. This is already quite a lot, because he seems to be on very friendly

---

[142] *Elektrotechnische Zeitschrift* **36**, 488 (1892).

[143] Scott Walter, Étienne Bolmont and André Coret (Eds.), *La Correspondance entre Henri Poincaré et les physiciens, chimistes et ingénieurs,* Birkhäuser, 2007.

terms with them". Angelo Battelli was also working on high-frequency alternating currents, so Albert obviously considered himself an expert in wireless telegraphy.

Giuseppe Jung's library

The fact that Michele worked with Albert on wireless telegraphy is evident, as we shall now see, from the very contents of his uncle Giuseppe Jung's library. First of all, let us look at the history of the library. In November 1926, Jung's youngest daughter, Maria Vittoria, donated her father's personal library to the Milan Polytechnic. Today it is divided between the Polytechnic's Francesco Brioschi mathematics library and the Giovanni Ricci library of the university's Mathematics Institute, where it appears in the 1935 register. The head of the Polytechnic's library, Prof. Umberto Cisotti, divided the donation according to the interests of the two institutions. This major donation, comprising around four hundred and fifty books and some two thousand five hundred issues (mainly offprints of articles), to which we will return later, is detailed in the library entry registers. A quick glance at the titles of the books shows that, in addition to his speciality in graphic statics and geometry, Jung was interested in the related fields of mechanics in general (solids, hydrodynamics, elasticity) and the various branches of analysis. The books are in French, German or Italian. Books on physics are rare: however, they include Johannes Diderik van der Waals' 1881 treatise on the theory of gases, James Clerk Maxwell's 1885-1887 French translation of his treatise on electricity and magnetism, several works by Gustav Kirchhoff and Max Abraham[144] and August Föppl's 1907-1908 theory of electricity. In the period from 1899 to 1901, which is of interest to us because of Michele and Albert's presence in Milan, there are a remarkable set of seven works on electricity and wireless telegraphy, all published between 1897 and 1901: Augusto Righi's 1897 optics book on electrical oscillations[145]; Oreste Murani's 1901 physics treatise[146]; Henri Poincaré's *Theory of potential* (1899)[147], as well as his work on Hertzian oscillations (1899)[148] and the second edition of *Electricity and Optics* (1901)[149]; André Broca's *Wireless*

---

[144] The German physicist Max Abraham, appointed professor at the Polytechnic in 1909 and a member of the Lombard Institute in 1911, was a colleague of Jung's until he left Milan in 1915 with Italy's entry into the war.

[145] Augusto Righi, *L'ottica delle oscillazioni elettriche*, Bologna, Zanichelli, 1897.

[146] Oreste Murani, *Trattato elementare di fisica*, Milan, Hoepli, 1901.

[147] Henri Poincaré, *Cinématique et mécanismes, potentiel et mécanique des fluides*, Paris, Carré & Naud, 1899.

[148] Henri Poincaré, *La Théorie de Maxwell et les oscillations hertziennes*, Paris, Carré & Naud, 1899.

[149] Henri Poincaré, *Électricité et optique*, Paris, Carré & Naud, 2nd edn, 1901.

telegraphy (1899)[150]; and a treatise on electricity and its applications by René Colson (1900)[151]. This collection is reminiscent of the updated bibliography of a thesis student working on wireless telegraphy around 1900[152].

These books are well known. Righi's book, which presents his various experiments, is considered to have had a certain influence on Marconi. Poincaré's 1899 work, without the calculations, repeated the content of the 1894 book on Hertzian oscillations mentioned above, updating it on the qualitative and experimental aspects. In the second edition of *Electricity and Optics*[153], he gave a detailed discussion of recent electrodynamic theories, in particular those of Lorentz (1895) and Larmor. Broca's work, based on the lectures he gave at the École Polytechnique, completes the subject from the point of view of electrical diagrams. It was recommended by André Blondel in his report on wireless telegraphy, presented at the International Electricity Congress held in Paris from August 18 to 25, 1900 during the Universal Exhibition. It should be added that the two works by Poincaré published by Carré & Naud and the one by Colson published by Gauthier-Villars bear the "Hommage des éditeurs" [tribute from the editors] mark. This shows that these were period works sent by the publishers, if not to Jung directly, at least to the professors at the Milan Polytechnic, or even to its director, Giuseppe Colombo. Jung, who seems to have had no professional interest in these subjects, could have obtained these works and made them available to his nephew.

A thesis by Michele in 1900-1901?

A note in the *Collected Papers* relates Albert's work on wireless telegraphy to Michele's work on the safety of power stations. This hypothesis is not inconsistent with the fact that Michele may have begun a thesis on the subject in parallel in 1900-1901. Michele's letter to Albert, January 17, 1928, shows us that he had indeed started a thesis and that it was an important link between them: "When we last saw each other, you twice mentioned attaining a doctoral degree, which in fact never happened — and I had to think a long time about how many quite different bonds there are between us[154]". We also know that Albert encouraged his friend to start a thesis at the same time as he did. On October 3, 1900,

---

[150] André Broca, *La Télégraphie sans fils*, Paris, Gauthier-Villars, 1899.

[151] René Colson, *Traité élémentaire d'électricité, avec principes et applications*, Paris, Gauthier-Villars, 1900.

[152] The presence of the last three works on physics published by Poincaré is all the more remarkable given that Giuseppe Jung, despite being a mathematician, had in his library only one offprint by Poincaré, on Fuchsian functions, out of a personal library of some two and a half thousand items.

[153] Poincaré deposited this new edition with the library of the Institut de France on December 14, 1900.

[154] Letter from Michele Besso in *CPAE (English translation supplement)*, vol. 16, Doc. 132, p. 141.

for example, he wrote to Mileva from Milan: "I prodded him very much to become a Dozent, but I doubt very much that he'll do it. He simply doesn't want to let himself and his family be supported by his father, this is after all quite natural. What a waste of his truly outstanding intelligence[155]".

Michele, who graduated from ETH in 1895, does not seem to have had the opportunity to start his thesis before then. Since he was six years older than Albert, he could not have started his thesis too late either and 1900 seemed to be the right year for him to do so. We saw in the third chapter that he had a somewhat protected job with the Barberis company that employed him. The interruption in his thesis may have been linked to his father's death in October 1901 and his move to Trieste. Although he began a thesis on wireless telegraphy in October 1900, there is currently no information to suggest which scientific group he might have planned to work with, but we can give a few clues. Apart from Righi in Bologna, who was a friend of his uncle Giuseppe Jung and his colleague at the Lombard Institute, Michele could have kept in touch in Rome, where his uncles Marco and Beniamino lived, with Pietro Blaserna, of whom he had been one of the best students and who was a friend of the family. In Blaserna's environment, Michele could count on his cousin Moisè Ascoli, the son of Graziadio, Marco Besso's uncle, originally from Gorizia near Trieste. After training with Giovanni Cantoni at the University of Pavia, where he was Felice Casorati's assistant in 1880, Moisè Ascoli became assistant professor of technical physics at the University of Rome in 1890-1891, the same year Michele was a student. He became a full professor in 1897 and gave a public lecture on wireless telegraphy in connection with Marconi's work. Between 1897 and 1902, he published a dozen articles on the subject[156]. He would have been an ideal supervisor for Michele's thesis. Quirino Majorana, Blaserna's assistant, was also working on the subject, as we saw earlier.

An example of following up on scientific questions: thermoelectricity

The letters to Mileva allow us to follow the origin and evolution of Albert's questions on a first subject, that of the specific heat of bodies and conduction, which also allows us to understand his scientific approach. In October 1899, Albert wrote to Mileva from Milan, on the subject of thermoelectricity:

> I have been studying here a lot & have completed my considerations about the study of the basic laws of thermoelectricity. I have also devised a method of great simplicity, which permits one to decide whether the latent

---

[155] Letter to Mileva Marić in *CPAE (English translation supplement)*, vol. 1, Doc. 79, p. 152.
[156] Antonio Casella and Guido Lucchini, *Graziadio e Moisè Ascoli. Scienza, cultura e politica nell'Italia liberale*, Pavia, La Goliardica Pavese, 2002.

heat in metals is to be reduced to the motion of ponderable matter or of electricity, *i.e.*, whether an electrically charged body has a different specific heat than an uncharged one. All these questions are connected with the analysis of the "thermo-element". The methods are very simple to carry out & do not require any equipment that is not readily available to us[157].

Over the next two years, he wondered about the precise nature of this *thermo-element* – molecule, resonator (oscillating electric dipole), free electric charge – and about the experiments that would make it possible to determine it. Thus, at the beginning of September 1900, in Milan, he told Mileva that "for the investigation of the Thomson effect [he has] again resorted to another method, which has some similarities with yours for the determination of the dep[endence] of K [thermal conductivity] on T [Temperature]" and enthused: "If only we could already start tomorrow! With Weber we must try to get on good terms at all costs, because his laboratory is the best and the best equipped[158]". Their work, initially carried out as part of their final year diploma, is linked to the Thomson effect, which reflects the appearance of a voltage between the ends of a junction formed by two metals of different types heated to different temperatures. We do not know the results of the experiments described above, but Albert wrote to Carl Seelig on April 8, 1952: "My wife's *Diplomarbeit* and mine too, related to heat conduction and were of no interest to me and deserved no consideration[159]".

The letters to Mileva also tell us that in September 1900 Albert had taken a great interest in reading Ludwig Boltzmann's lectures published in the two volumes of his *Lectures on gas theory* of 1896 and 1898. He even wrote to her that he had "studied the entire Boltzmann", including "a part of spherical harmonics, in which I have now even got quite interested", useful for dealing with transport phenomena. He went into more detail on this subject in Eduard Heine's book, to which Boltzmann referred and which he would recommend to Mileva a few months later. Heine's book on spherical harmonics may have been borrowed for him by Michele from his uncle Giuseppe Jung[160]. Although

---

[157] Letter to Mileva Marić in *CPAE (English translation supplement)*, vol. 1, Doc. 58, p. 136.

[158] Letter to Mileva Marić in *CPAE (English translation supplement)*, vol. 1, Doc. 74, pp. 147-148.

[159] Letter Hs_304_14, Albert Einstein online archive at ETH Zurich. It is true that this work had earned them, along with Weber, marks of 16/24 (Mileva) and 18/24 (Albert), lower than those of their three other classmates who had graduated in mathematics.

[160] Eduard Heine, *Handbuch der Kugelfunctionen, Theorie und Anwendungen*, Berlin, Reimer, 1878 (vol. 1) and 1881 (vol. 2). Heine's volume can be found both in the library of the Brera Astronomical Observatory and in Jung's library. Albert could have gained access to the observatory's library by applying directly to its director, the famous astronomer Giovanni Schiaparelli until 1900, or his successor Giovanni Celoria. But he showed no interest in astronomical subjects in his letters, and the observatory library did not hold the physics journals that interested him. Heine's work may well have been passed on to Albert by Michele.

Albert probably did not read all of Boltzmann's five hundred pages[161] in detail, he was probably convinced by its theoretical approach, which based the thermodynamic properties of gases and liquids on the dynamics of molecules.

However, at the end of March 1901, the subject of the nature of the *thermo-element* returned to the forefront of his mind, probably following his reading of Max Planck's articles on blackbody radiation. On the train to Milan, he wrote to Mileva on March 23 that "An original idea occurred to me on the trip. It seems to me that it is not out of the question that the latent kinetic energy of heat in solids and liquids can be conceived of as the energy of electrical resonators[162]". "In this case", he adds, "the specific heat and the absorption spectrum would have to be interrelated". He asked Mileva to investigate any difference between the specific heats of opaque and transparent bodies, expecting a deviation from Dulong and Petit's law for glass (whose resonators are not excited because it is transparent), and then suggested that she should analyse the metallic reflection of light, in order to determine the absorption coefficients and compare their temperature dependence with that of the specific heats.

A few days later, however, he doubted Planck's resonators and it was Drude's electron theory, read in the library of the Lombard Institute, "a kinetic theory of electric and thermal phenomena in metals, entirely in the spirit of the kinetic theory of gases", that was now in his favour, when he wrote to Mileva on April 4:

> On the other hand, I have in my hands a study by Paul Drude on the electron theory, which is written to my heart's desire, even though it contains some very sloppy things. Drude is a man of genius, there is no doubt about that. He also assumes that it is mainly the negative electric nuclei without ponderable mass which determine the thermal and electric phenomena in metals, exactly as it occurred to me shortly before my departure from Zurich[163].

However, a month later, his reading of an article by Max Reinganum, which attributed a finite mass and negligible size to electrons and explained that they could not be responsible for the specific heat of metals, probably led Albert to write to Drude "to draw his attention to his mistakes[164]" and formulate "two objections to his electron theory". He interpreted the reply from Drude, director of the Göttingen physics laboratory and Editor-in-Chief of the *Annalen der Physik*, which did not even discuss his arguments, as an authoritative response and "an irrefutable evidence of its writer's wretchedness". He wrote to Mileva from Winterthur in July: "From now on I'll not turn

---

[161] This was demonstrated by his return to the foundations of statistical thermodynamics between 1902 and 1904, in his three articles published in the *Annalen der Physik*.
[162] Letter to Mileva Marić in *CPAE (English translation supplement)*, vol. 1, Doc. 93, p. 159.
[163] Letter to Mileva Marić in *CPAE (English translation supplement)*, vol. 1, Doc. 96, p. 162.
[164] Letter to Mileva Marić in *CPAE (English translation supplement)*, vol. 1, Doc. 110, p. 173.

any longer to this kind of person but will rather attack them mercilessly *via* journals, as they deserve[165]". Albert was deeply distressed to see any possibility of a physics teaching post in Germany closed in this way, especially as Drude had told him that he had informed "another (infallible) colleague [who] shares his opinion[166]". Albert did not return to the subject until after 1905.

The letters to Mileva, most of them written from Milan, bear witness to the scientific passion and maturity of the young Albert, who was full of scientific proposals to be submitted to the verdict of experiment or to be reconsidered theoretically. It also appears – as is natural for a student preparing for a degree or a thesis – that his ideas were linked to a great deal of reading, which he approached with enthusiasm, but also critically, as we have seen from the example of books or articles by Boltzmann, Planck or Drude, underlining the importance of his bibliographical work in reference journals and manuals.

Albert's professional worries

Links with his father's new company business

In Milan, Albert could not ignore his family's professional environment. Hermann, his father, and his new company set up an electricity production plant at Canneto sull'Oglio in Mantua in 1899 and another at Isola della Scala in the municipality of Verona in April 1900. He received money from these municipalities for the use of electricity in the form of an annual rent[167]. Maja Einstein recalls her father's last industrial activity as follows: "With money provided by relatives, Hermann Einstein then turned to installing power stations, supplying whole villages with lighting. This time, success seemed to be his[168]".

Hermann planned to take his son, a recent ETH graduate, to visit them in September 1900. Albert naturally envisaged having to take over from his father and agreed to go to the field. In mid-August 1900, he wrote to Mileva:

> I could have had a position in life insurance for 3 weeks for 8 fr. [Francs] per day, but I declined because I thought that I could use my vacations better by studying something proper & then in Italy learning the trade of

---

[165] Letter to Mileva Marić in *CPAE (English translation supplement)*, vol. 1, Doc. 114, p. 176.

[166] Letter to Jost Winteler in *CPAE (English translation supplement)*, Doc. 115, p. 176. Jürgen Renn specifies that Drude must in fact be referring to Ludwig Boltzmann: see Renn Jürgen, "Einstein's controversy with Drude and the origin of statistical mechanics: A new glimpse from the 'Love Letters'", *Archive for the history of exact sciences* **51**, 315-354 (1997).

[167] Fabio Bevilacqua, *Albert Einstein, ingegnere dell'universo, op. cit.*

[168] "Albert Einstein - A Biographical Sketch" (Translated Excerpts) by Maja Winteler-Einstein in *CPAE (English translation supplement)*, vol. 1, pp. xvii.

my father. After all, it could happen that he suddenly fell ill or would be otherwise engaged & he has nobody at his disposal[169].

A few days later, he confirmed that his father's business had improved: "My father has become a completely different man now that he no longer has to worry about money". The departure took place the following week: "Friday. So, tomorrow we are leaving for the trip, but we will be back in a week, so just keep sending me your little letters. Luigi Ansbacher might come to visit us".

Shortly afterwards, in October 1902, at the age of twenty-three, Albert had to make a decisive choice. His father had just died in Milan. For several months, Albert had been working as a technical expert at the Patent and Intellectual Property Office in Berne[170]. He preferred to stay on as an employee rather than take over his father's business as an engineer. It is likely that the difficulties encountered by his father during his entrepreneurial career did not encourage him to take over the business in order to maintain an industrial activity in an extremely competitive environment that requires constant investment, all the more so as he wished to pursue his own scientific and academic goals. Rudolph Einstein[171], Hermann's cousin, who had partly financed the company, appointed the young Milanese lawyer Luigi Ansbacher to manage it[172].

Looking for a position as a university assistant

Albert began his first thesis at the end of 1900 on molecular forces, to which we will return in the next chapter. His colleagues Ehrat, Grossmann and Kollros, who had graduated from ETH with a degree in mathematics, all obtained assistant posts with one of their professors. Only Albert, who had a degree in physics, remained without a post. He deduced that Weber was not supporting him among his colleagues, and even worse, began to think that he was trying to block his application. In September 1900, he wrote to his mathematics professor, Adolf Hurwitz, to apply for a position as his assistant, a possibility that was briefly discussed but never confirmed. In December 1900,

---

[169] Letter to Mileva Marić in *CPAE (English translation supplement)*, vol. 1, Doc. 72, pp. 145-146.

[170] It should be noted in passing that the filing of patents, which had recently been introduced in European industry (the Federal Patent Office in Berne had been set up in 1888), was an essential part of a company's survival and that Jakob Einstein, Albert's uncle, had himself filed seven patents. Albert was therefore made aware of this need at a very early age, which indirectly prepared him for his current position in Berne.

[171] He was also the father of Elsa, Albert's wife in 1919. Elsa's mother is Fanny Koch, Albert's mother's sister.

[172] What became of the company? There are only a few references to Luigi Ansbacher in the letters exchanged between 1903 and 1912 between Albert and Prof. Alfred Stern of Zurich, to whom the Ansbacher family had introduced him, which appear in: *CPAE (English translation supplement)*, vol. 5: The Swiss Years: Correspondence, 1902-1914, translated by Anna Beck, Princeton, Princeton University Press, 1995.

he sent Ludwig Boltzmann a personal copy of his article on capillarity, as Mileva pointed out to her friend Helene Savić:

> Albert has written a paper in physics that will probably be published very soon in the physics *Annalen*. You can imagine how proud I am of my darling. This is not just an everyday paper, but a very significant one, it deals with the theory of liquids. We sent also a private copy to Boltzmann and would like to know what he thinks about it, let's hope he is going to write to us[173].

He probably also made an unsolicited application to Boltzmann on this occasion, since his article served as his "visiting card" to introduce himself. In the autumn of 1900, Albert went to Vienna (where Boltzmann taught until 1900) to apply for a "practical job", while Mileva unsuccessfully sought a position at a girls' grammar school. He then applied to Riecke in Göttingen, whom he knew from his articles on electron theory and later to Drude, as we shall see, and also sent unsolicited applications in March to Otto Wiener and in April, from Milan, to Wilhelm Ostwald and Heike Kamerlingh Onnes, in whose laboratory Max Reinganum was working, as mentioned earlier. Ostwald's candidacy was clearly close to his heart, as his father, Hermann, wrote independently on April 13 to support his son. This interest in Ostwald should be seen in the light of the enthusiasm Albert shared with Mileva about his reading in physical chemistry, in particular the theory of ions in electrolytes, on which the theory of electrons was based to describe conduction in metals. He also sent his article on capillarity to Giuseppe Jung, who, as we have seen, took it upon himself to contact his colleagues Righi and Battelli on his behalf.

In May 1901, in desperation, he thought of asking Michele's father once again to enter the insurance business, as he had already envisaged on leaving ETH. Mathematical (statistical) skills are particularly valued in actuarial work. He wrote to Mileva at the end of May: "I have already wondered whether old Besso couldn't find a job for me in insurance. After all, he is the general manager of a company[174]". He applied unsuccessfully for a position in the technical department of the cantonal school in Burgdorf, despite the support of Jost Winteler, who had put him up in Aarau. From mid-May to mid-July 1901, he replaced Jakob Rebstein (1868-1951) at the Winterthur Technical School, a position to which he had been invited, much to his surprise, without asking. Albert told Mileva: "This Rebstein is probably Hertzog's former assistant, whom we knew, actually". Rebstein used to teach them mechanical engineering at ETH. He was also the son of Jakob Johann Rebstein (1840-1907), whose optional courses Albert had taken at ETH in the second half of 1897-1898 on the mathematical foundations of statistics and personal insurance. Professor Becker's tribute to him on his death

---

[173] Letter from Mileva Marić to Helene Savić in *CPAE (English translation supplement)*, vol. 1, Doc. 85, pp. 155-156.

[174] Letter to Mileva Marić in *CPAE (English translation supplement)*, vol. 1, Doc. 111, p. 174.

states that Jakob Johann also acted as an advisor to Swiss companies[175]. It is therefore conceivable that he was linked to Giuseppe Besso, who ran a subsidiary of General Insurance until 1879 in Zurich. Jakob Johann had been a teacher at the Frauenfeld cantonal school from 1861 to 1877, where his pupil was undoubtedly Jost Winteler, who graduated from the school in 1866 and could also have been involved with Rebstein.

Albert then stayed in Switzerland, where he became a teacher at Dr Nuesch's public school in Schaffhausen, preparing a young English student, Louis Cahen, for the ETH entrance exam. Dr Nuesch is best known for his discovery of a Palaeolithic cave in this region. Today, visitors to the town museum are greeted by a large-format photograph of Nuesch's archaeological excavations. But the material situation did not suit Albert, as his letters to Mileva show, and his relationship with Nuesch became stormy. The bright spot came in the form of his friend Marcel Grossmann, whose father interceded on his behalf to obtain a position at the Federal Intellectual Property Office in Bern, with its director, Friedrich Haller. Albert therefore resigned his teaching post in Schaffhausen and moved to Bern in February 1902, where he began work in June. At the beginning of February 1902, Mileva gave birth to a baby girl, Liserl, of whom history has no record. At this point, Albert asked to be reimbursed for his thesis fees at Zurich University. Another stage in his life began. But let's return to his environment in Milan, this time from the point of view of social relations.

Scientific protection

Albert wrote to Mileva in the spring of 1901, as we have already mentioned: "I have quite good connections here. That is to say, Mr. Ansbacher is a close friend of the professor of chemistry at the Milan Polytechnicum, and, further, Michele's uncle is a professor of mathematics[176]". We shall see that Albert remained in contact with Michele's uncle for several years and that the chemistry professor in question was probably Luigi Gabba.

Albert's lasting link with Giuseppe Jung through his personal library

Let us go back to the contents of Giuseppe Jung's personal library, which we studied above with regard to his few physics books and turn to the offprints of articles. As you might expect, their list contains only mathematical articles,

---

[175] F. Becker, "Professor Dr. Jakob Rebstein", in *Beilage "Nekrologe" zu den Verhandlungen der schweiz. naturf. Gesellschaft*, Freiburg, 1907. Communication ETH Zurich Archives.
[176] Letter to Mileva Marić in *CPAE (English translation supplement)*, vol. 1, Doc. 94, pp. 160-161.

but with one notable exception: all the offprints of Albert Einstein's publications from 1901 to 1906, a total of twelve articles[177]. Until recently, there was no trace of an ongoing exchange between Albert and Jung, beyond Albert's first article in 1901. Perhaps Albert was still hoping to get a university post in Italy before he obtained the permission to give lectures at the University of Bern in 1908 and continued to maintain links with Jung? Or was it part of the personal collection that Michele had mislaid in 1919, as we said in the introduction? Unfortunately, according to information gathered indirectly from a former head of the library at the Milan Polytechnic, it was decided in the 1980s to send these articles to the shredder because of their poor state of preservation. It is therefore impossible to know whether they contained any annotations. Jung's protection would thus have consisted of giving Albert access to the library of the Lombard Institute, providing him with books through Michele (Heine, wireless telegraphy, etc.) and interceding on his behalf with his physicist colleagues (Righi, Battelli, etc.) for an assistant post.

The professor of chemistry at the Polytechnic Institute

So, who was the "professor of chemistry" at the Polytechnic mentioned above, who guaranteed Albert his protection through Bernardo Ansbacher? The *Collected Papers* indicate four possibilities: Pietro Gorbetta, Luigi Gabba, Guglielmo Koerner and Angelo Menozzi, who was Koerner's student according to the Polytechnic syllabus in 1901. In 1900, Gabba and Koerner were the most renowned figures. They founded the Milan Chemical Society in 1895, of which Gabba was elected Vice-President, then president following Koerner. This society was the prelude to the Italian Chemical Society.

It seems to us that Luigi Gabba (1841-1916) is the scientist of choice. He had held a professorship in chemistry at the Milan Polytechnic since 1872, where he taught "general and technological chemistry". He had attended the Society for the Encouragement of Arts and Crafts, where Colombo taught, before studying at the University of Pavia. He graduated in physical and chemical sciences from the University of Pisa in 1866, where he studied at the same time as Davide Besso, Michele's mathematician uncle. After enlisting with the Garibaldi volunteers, he specialised in chemistry in Berlin. He became a member of the Lombard Institute in 1877. In 1883, when a new chair of chemistry was created at the Polytechnic, he opted for the chair of technological chemistry, and at the same time directed the school's chemistry laboratory[178]. In his

---

[177] Ref. 3400 and 2766 to 2776 from the 1926 entry register of the Mathematics Library of the Milan Polytechnic. Einstein's articles complete the donation of Jung's fascicles (with the exception of the one given by Michele, which is separated from the others in the list).

[178] Gian Piero Marchese, "Luigi Angelo Gabba", in *Dizionario biografico degli Italiani*, vol. 50, 1998.

teaching, the emphasis was on experimental aspects, in line with the practice of many professors inspired by German methods. Luigi Gabba co-founded the *Gazzetta chimica italiana* in 1870 and published a treatise and textbook on chemistry with the publisher Hoepli. He also translated English and German works into Italian. He was also the first adviser to the board of the Italian Alpine Club (CAI), whose library today bears his name.

The link with Bernardo Ansbacher may also involve his elder brother, Carlo Francesco Gabba (1835-1920), a renowned specialist in public law and legal philosophy, who taught in Pisa in 1862 and then in Turin. He represented Italy at the international congress on the protection of industrial property in Brussels in 1897. He was also a corresponding member of the Lombard Institute, in the Letters and Moral and Political Sciences section. This link could also involve his younger brother, Bassano Gabba (1844-1928), also a lawyer and at the time a correspondent member of the Lombard Institute and Milan city councillor responsible for public education.

The Ansbacher family and the musical environment in Milan

The lawyer Bernardo Ansbacher (1845-1914), Luigi's father who took over the Einstein business after Hermann's death, is mentioned as a family friend in the *Collected Papers*. The Savallo guide to Milan states that Bernardo was a member of the Milan Quartet Society, an elitist society founded in 1864. Its aim was to promote chamber music and introduce the public to works by Austrian classical composers Mozart, Haydn and Schubert, among others, as well as German composers. Major concerts were held under the aegis of this Society, such as those conducted by Arturo Toscanini and Richard Strauss in the early 1900s.

Albert Einstein was an excellent violinist, as is well-known. He said he learnt to play the violin between the ages of six and fourteen, but only really progressed at the age of thirteen, when he discovered Mozart's sonatas. His sister Maja recalls that "Music was played often and well at home". An examination report dated March 31, 1896 from the cantonal school in Aarau confirms this:

> The violin playing still revealed some stiffness in bowing techniques here and there, but otherwise the results were quite gratifying with regard to technique as well as with regard to intonation. One student, by name of Einstein, even sparkled by rendering an adagio from a Beethoven sonata with deep understanding[179].

Albert also played in Italy, not only in Pavia but also in Milan. In March 1897, his mother wrote to Marie Winteler, one of the daughters of the family

---

[179] Inspector's report on a music examination, Aargau kantonsschule in *CPAE (English translation supplement)*, vol. 1, Doc. 17, p. 12.

where Albert had stayed in Aarau – his youthful flirt: "Each day I make plans to write you, but when Albert is here I do not accomplish anything: There is so much laughing, joking, and music-making going on that there is not enough time for anything else[180]". Later, in September 1900, Albert wrote to Mileva: "Luigi Ansbacher, about whom we are always teasing Maja, is going to arrive in a few days – I am looking forward to playing music with him". Luigi enrolled in his first year of law studies at the University of Pavia in 1895, graduating in 1900[181]. He was a friend of Arturo Toscanini, with whom he appears in a photograph in the history section of the Milan Quartet Society's online site. We have seen other guests come to the Einsteins' house on occasion to "make a little music", such as Giovanni Barberis, director of the Society for the Development of Italian Electrical Companies, where Michele worked. It's worth noting that the young Albert attached great importance to this musical environment, which enabled him to take his mind off things in Milan. For example, before sharing his final thoughts on molecular forces with his friend Marcel Grossmann on April 14, 1901 in Milan, he admitted that his "musical acquaintances protect me here from getting sour".

And it's not just music that allows him to escape. He was also particularly fond of alpine excursions. In July 1899, he climbed the Säntis, east of Zurich and south of St Gallen, in September 1900 a mountain on the shores of Lake Maggiore, spent the summer on the Simplon, on the Milan-Berne railway line and in July 1901 climbed the Klausen near Altdorf, on the Milan-Zurich line. He also walked from Voghera to Genoa, crossing the Alpilles, to join his uncle Jakob Koch, a brother of his mother who was a cereal importer there. We should also remember that he spent the summer of 1895 in Airolo on the St Gotthard massif, the same year that the CAI [Italian Alpin Club] graduated its first mountain guides there, and that inaugural hikes must have been organised for the occasion. The Milan section of the CAI, founded in 1873 and based in the same building as the Polytechnic, had a membership of five hundred and eighty, including a number of prominent Milanese figures we met, such as Luigi Gabba, Ulrico Hoepli and even Carlo Monti. Chamber music and hiking were the favourite pastimes of high society, offering a privileged and informal way to socialise.

---

[180] Letter from Pauline Einstein to Marie Winteler in *CPAE (English translation supplement)*, vol. 1, Doc. 32, p. 31.
[181] *Annuario della R. Università di Pavia, anno scolastico 1895-1896* et *1900-1901*, Pavia, Bizzoni, 1896 and 1900.

6

Three Albert's scientific questionings (1898-1901) in connection with his later 1905 work

In this chapter, we place three important works by Albert Einstein in the scientific context we have been studying. Starting with a theme that dates back to 1898, the relative motion of matter with respect to the ether, we will give Albert and Michele Besso's views on Mach's influence and discuss the experiments that Albert devised in 1901 to demonstrate the motion of the Earth in the ether. This attempt is undoubtedly one of the "bad leads" mentioned in retrospect by Michele. We will then discuss Albert's first thesis on molecular forces, which he extended to weakly compressed gases in April 1901 before abandoning it a few months later, at the beginning of 1902, preferring to devote himself to the kinetic theory of thermal equilibrium. The discussion will be conducted in relation to Max Reinganum's article, published in the *Jubilee* for Lorentz, which is in the library of the Lombard Institute where Albert works. Another "bad lead". Finally, we will confirm Jürgen Renn's hypothesis that Albert had his first idea of light quanta as early as 1901, by placing ourselves in the perspective of his readings in Milan at the library of the Lombard Institute.

The recurrent problem of the relative motion of matter and ether (1898-1901)

Reading Ernst Mach

We recalled in the introduction that Albert first raised questions about relativity in 1898. It should be remembered that this was obviously not "relativity" in the sense of the 1905 theory of special relativity, in which the ether was explicitly rejected as an absolute reference frame to which motion of bodies could be related. In fact, for him at the time it was simply a matter of discussing the "bodies' relative motion with respect to the luminiferous ether", mentioned in his letters to Mileva in September 1899 and which was still the first subject of his discussions with Michele in April 1901, in the form of "the fundamental separation of luminiferous ether and matter". We saw in the introduction that Michele referred to 1898 as an important year in which he shared Albert's thoughts: it was the year in which he introduced him to the philosophical ideas of Ernst Mach.

The Einstein-Besso correspondence provides us with some retrospective information on their respective perceptions of Mach's role in the construction of the theory of relativity. The first discussion arose in 1917. Albert had just criticised a text on relativity by their mutual friend, Friedrich Adler, "in which he presents extremely expansively and with prophetic conviction quite worthless hypercritical points[182]" and in which "he rides and spurs Mach's hack to exhaustion". Michele is much less categorical than Albert and replies: "As regards Mach's little steed, let us not berate it so much; did it not serve trustily in the hellish ride through the relativities? And who knows whether it does not carry its rider, Don Quixote de la Einsta, through the nasty quanta as well![183]". In Michele's eyes, reading Mach therefore fuelled their first scientific discussions on relativity. However, the disagreement between the two friends over its real role would persist for a long time to come. It can be explained by the fact that Albert took a physicist's point of view, to which Mach's analysis did not provide any concrete elements for the construction of a theory, whereas Michele took a philosophical point of view.

The discussion resumed in 1947, on the subject of information given by Michele to which we referred in the introduction and which was clearly not to Albert's liking. Michele wrote as a preamble to a letter that had taken him almost two months to write: "With what follows, I imagine I am fulfilling a duty. If you think otherwise, don't hold it against the man who is still in his 75th year". Then he picks up on some of Albert's remarks that have affected him: "My dear, best and old friend, you told me not to worry about the inaccurate information I gave about you, giving as the reason that people have lied so much about you that it really doesn't matter". Challenging him again on the need to make the genesis of his scientific ideas, a subject he had been raising in his letters for over twenty years as we have seen, he continued on Mach:

> As far as the history of science is concerned, Mach is, in my view, at the centre of developments over the last 50 or 60 years. Is it true
>
> - that a slightly older engineer [Michele himself], around 1897 or 1898, sent back to Mach the young interested party [Albert], who was passionate

---

[182] Letter to Michele Besso in *CPAE (English translation supplement)*, vol. 8, Doc. 331, pp. 321-322.

[183] Letter from Michele Besso in *CPAE (English translation supplement)*, vol. 8, Doc. 334, p. 324.

about science and who was fascinated by the question of the existence of the ether and atoms

- to Mach, which had been pointed out to the engineer by the famous Slovakian professor [Aurel Stodola] at the ETH; and that this reference took place during the young physicist's training, at a time when Mach's teaching referred decisively to observable phenomena – and perhaps indirectly, precisely, to clocks and measurements?[184]

In his book entitled *The science of Mechanics: a critical and historical account of its development*, published in 1883, Mach challenged Newton's notion of an absolute frame of reference and gave meaning only to the relative motion of bodies. However, he did not rule out the possibility that a body's motion could be determined by the environment in which it moved, and that this environment could therefore replace absolute space. He even added that "from the point of view of science it would be in every respect a more valuable acquisition than the forlorn idea of absolute space", and Albert could logically have initially placed himself in this perspective. This is also apparent from his autobiography of 1949. After describing himself on a philosophical level and evoking "Mach's greatness in his incorruptible scepticism and independence", he immediately adds: "in my younger years, however, Mach's epistemological position also influenced me very greatly, a position which today appears to me to be essentially untenable[185]". This statement relates not only to Mach's critical attitude towards the existence of atoms, but perhaps also to the relationship between matter and the ether.

An attempt to demonstrate the relative motion of matter with respect to the ether

From 1899 onwards, Albert thought about designing experiments to detect the supposed motion of matter with respect to the ether. On September 10, he reported to Mileva: "A good idea occurred to me in Aarau about a way of investigating how the bodies' relative motion with respect to the luminiferous ether affects the velocity of propagation of light in transparent bodies[186]". At the end of September, he added that he "also wrote to Professor Wien in Aachen about the paper on the relative motion of the luminiferous ether against ponderable matter, which the "Principal [Weber]" treated in such a stepmotherly fashion". Two years later, he was working

---

[184] *Einstein Albert, Besso Michele, Correspondance (1903-1955), op. cit.,* (B. 74) p. 228. Translation from French.

[185] Albert Einstein; Notes for an autobiography, *The Saturday review of literature,* November 26 (1949), p. 11.

[186] Letter to Mileva Marić in *CPAE (English translation supplement),* vol. 1, Doc. 54, p. 132.

harder than ever on the subject, and in September 1901 he wrote to his friend Marcel Grossmann:

> A considerably simpler method of investigating the relative motion of matter with respect to luminiferous ether that is based on ordinary interference experiments has just sprung to my mind[187].

On December 17, he wrote to his future wife:

> I am now working very eagerly on an electrodynamics of moving bodies, which promises to become a capital paper. I wrote to you that I doubted the correctness of the ideas about relative motion. But my doubts were based solely on a simple mathematical error. Now I believe in it more than ever. Since that bore Kleiner hasn't answered yet, I am going to drop in on him on Thursday. I want to induce him at all costs to let me work during the Christmas vacation. I'll see whether I'll succeed in that[188].

Alfred Kleiner (1849-1916), professor of physics at the University of Zurich, his thesis supervisor, whom he saw two days later, encouraged him in this direction, according to what Albert told Mileva:

> He advised me to publish my ideas about the electromagnetic theory of light of moving bodies together with the experimental method. He thought that the experimental method proposed by me is the simplest and most appropriate one conceivable. I was very pleased with the success. I shall certainly write the paper in the coming weeks[189].

It seems that this article never saw the light of day. As for the experiments designed between September and December 1901, they can be compared with the one Albert described at Kyoto University in December 1922.

Kyoto 1922 memories

Albert was in Japan at the time and was officially informed that he had won the Nobel Prize in Physics, after which he was invited to give an impromptu lecture. He devoted it to the genesis of his ideas on relativity. This reminiscence is very interesting, as it shows him, for the first and only time in his life, making an explicit and relatively precise reference to his pre-1905 ideas. He explains: "Then I myself wanted to verify the flow of the ether with respect to the Earth, in other words, the motion of the Earth. When I first thought about this problem, I did not doubt the existence of the ether or

---

[187] Letter to Marcel Grossmann in *CPAE (English translation supplement)*, vol. 1, Doc. 122, p. 180.

[188] Letter to Mileva Marić in *CPAE (English translation supplement)*, vol. 1, Doc. 128, pp. 186-187.

[189] Letter to Mileva Marić in *CPAE (English translation supplement)*, vol. 1, Doc. 130, pp. 188-189.

the motion of the Earth through it". He went on to describe the experimental set-up he had in mind:

> I thought of the following experiment using two thermocouples: set up mirrors so that the light from a single source is to be reflected in two different directions, one parallel to the motion of the Earth and the other antiparallel. If we assume that there is an energy difference between the two reflected beams, we can measure the difference in the generated heat using two thermocouples. Although the idea of this experiment is very similar to that of Michelson, I did not put this experiment to the test[190].

He was therefore looking for a variation in the relative intensity of light, proportional to the ratio V/c of the speed of the Earth to that of light, with an amplitude of 10^{-4}. From a theoretical point of view, such a possibility was doomed to failure, since it ran counter to the principle of relativity that would be enunciated in 1905. In 1901, however, it was still relevant in the context of comments on Lorentz's electrodynamic theory of 1895. Examples of this can be found in *L'Éclairage électrique* [*Electric lighting*], written by Alfred Liénard in 1898[191], or even in the second edition of *Electricity and Optics* by Henri Poincaré (in Jung's library), where the authors set out the transformation laws for electromagnetic energy quantities. In Chapter 6 of his book, Poincaré states a "theorem" according to which "the motion of the Earth has no influence on optical phenomena if we neglect the squares of the aberration $[V^2/c^2]$", and predicts, like Lorentz, that in an interference experiment the position of the fringes is unchanged, but he is rather ambiguous about their intensity: "If we could measure this variation in intensity, we would have a means of detecting the motion of the earth by optical phenomena. But we shall see that this is not the case: it is materially impossible to measure such a variation in intensity". He calculates that the energy of the radiation is transformed as $1 - V/c$ and, taking into account the direction in which the radiation is emitted, that a relative variation in intensity of $2V/c$, or 2×10^{-4}, can be expected. However, he adds, this value is far too small to be measured on photographic plates.

From an experimental point of view, Albert knew that he could take advantage of Weber's cutting-edge physics laboratory in Zurich. He was also familiar with *thermocouples* and their superior sensitivity to photographic plates. Thermocouples are based on the Thomson effect, which was mentioned in the previous chapter as part of the subject of Albert and Mileva's

---

[190] Albert Einstein, "How I Created the Theory of Relativity", translated by Yoshimasa Ono from notes in Japanese by J. Ishiwara, *Physics Today* **35** (8), 45-47 (1982). This was "an impromptu speech to students and faculty members, made in response to a request by K. Nishida, professor of philosophy at Kyoto University".

[191] Alfred Liénard's articles, in *L'Éclairage électrique* **16**, 320-334, 360-365 (1898) are discussed in Olivier Darrigol's article, "Poincaré, Einstein et l'inertie de l'énergie", *Comptes rendus de l'Académie des sciences* **1** (1), 143-153 (2000).

final year diploma. By assembling several hundred thermocouples in series, we obtain a *thermopile* whose electrical voltage can be measured by very sensitive equipment. The first thermopile was made by Leopoldo Nobili (1784-1835) and then popularised by Macedonio Melloni (1798-1854), so much so that at the time it appeared in every catalogue of laboratory equipment for measuring infrared radiation. Note that the term "thermopile" replaces "thermocouple" in the new translation of the text of the Kyoto conference in the *Collected Papers*[192], which seems to us to be more in line with reality. Perhaps Mileva is working on thermopiles when she redoes her final year at ETH and Albert sends her back to reading Heine on heat transport in cylinders? Thermopiles are generally cylindrical in shape.

In October 1901, two months before discussing his experiments and his "capital" paper on relative motion with Kleiner, Einstein was in Schaffhausen, where he met Paul Habicht. We saw in the introduction that he refers to him in connection with "a small electrostatic influence machine [...] which I used for static measurement of low voltages by multiplication". Should the idea behind the design of this instrument be linked to the use of thermopiles in the experiment described above? It is tempting to think so. The experiment Albert wanted to carry out required great sensitivity. The Einstein-Habicht electrometer, which came onto the market in 1908 after a gestation period of several years and numerous difficulties, finally achieved a sensitivity of 5×10^{-4}, quite close to the level that might have been expected in December 1901. Attempting such an experiment, contrary to the postulate of relativity, would have been one of the "bad leads" mentioned by Michele Besso.

The abandoned thesis on molecular forces (October 1900-December 1901)

Capillary phenomena and molecular forces in liquids

Capillarity is a subject mathematised by the work of Pierre-Simon de Laplace (1749-1827) and Carl Friedrich Gauss (1777-1855) at the beginning of the 19th Century. These authors defined the cohesive energy of a fluid based on the force of interaction between elements of infinitesimal volume. From a mechanical point of view, the interaction forces exerted on a molecule inside the fluid, summed in all directions, are in equilibrium, but this is not the case for molecules near the surface. The resulting force is then responsible for the surface tension of the liquid, a property used, for example, by insects to move

---

[192] *CPAE*, Diana Kormos Buchwald, József Illy, Ze'ev Rosenkranz, Tilman Sauer (Eds.), vol. 13: *The Berlin Years: Writings & Correspondence, January 1922-March 1923*, Princeton, Princeton University Press, 2012, Doc. 399, pp. 636-641 (English translation, see note 1).

on water or by soap bubbles to form. Capillarity entered the field of thermo-dynamics with the work of William Thomson (Lord Kelvin) in 1858, when he subjected the surface of the fluid to a cycle of transformations and defined the corresponding quantity of heat. The subject was more topical than ever at the end of the 19th Century. In 1897, Hermann Minkowski gave a course on capillarity at ETH, which Albert attended, as we mentioned in our discussion of Carlo Marangoni in Chapter 4. In 1907, in Sommerfeld's *Enzyklopädie der mathematischen Wissenschaften,* Minkowski devoted an article of almost sixty pages to this subject and gave a complete bibliography on it, including recent works by Josiah Willard Gibbs, *On the Equilibrium of Heterogeneous Substances* (in its German translation by Ostwald in 1892), Johannes Diderik van der Waals' *The continuity of the liquid and gaseous states of matter* (in its 1899 edition), Franz Neumann's *Lectures on the theory of capillarity* (1894), and Henri Poincaré's *Lessons on capillarity* (1895). These books appear in the ETH library and are therefore probably among the recommended reading for students in the mathematics and physics class.

Albert's first article, published in the *Annalen der Physik,* was entitled "Conclusions drawn from the phenomena of capillarity". It was submitted in December 1900 and published in March 1901. Using Clapeyron's relations between the calorimetric coefficients, he established by thermodynamic analysis the expression for the surface energy of a liquid in the form, $\gamma - T d\gamma / dT$ where T is the temperature and γ the surface tension of the liquid. The second term of this equation is introduced in Thomson's paper and recalled in Minkowski's course. Historians John Murrell and Nicole Grobert attribute the idea of adding a molecular contribution (γ-term) for the surface energy to Roberto Schiff[193], cited by Einstein for his experimental data, discussed in Chapter 4. The above equation is also found implicitly in Gibbs' book on heterogeneous substances. The historian of thermodynamics John Rowlinson sums it up as follows:

> In this paper he derived one equation that was later designated 'Einstein's equation' by Prigogine and his colleagues but which follows at once from the extension to surfaces of the Gibbs-Helmholtz equation of classical thermodynamics[194].

The originality of the article therefore does not lie in this equation, but rather in his approach of using it to determine the characteristics of atomic forces. Albert uses a mechanical determination of surface energy, which he equates with the thermodynamic determination. Instead of presenting this energy directly as an integral relating to elementary volumes of fluid, he

---

[193] John N. Murrell, Nicole Grobert, "The Centenary of Einstein's First Scientific Paper", *Notes and Records of the Royal Society Journal of the History of Science* **56** (1), 89-94 (2002).
[194] John S. Rowlinson, "Einstein: the Classical Physicist", *Notes and Records of the Royal Society Journal of the History of Science* **59** (3), 255-271 (2005).

focused on the forces between particles, as Boltzmann had done in his lessons on the theory of gases, which Albert read as early as 1899. Like Boltzmann, he considered the same type of force law, but assumed that it involved not the masses of the particles (as in the law of gravitation Gmm'/r^2), but coefficients "c_α" corresponding to the intrinsic characteristics of the chemical elements making up these particles. He makes the simplifying assumption that, apart from these specific coefficients, the interaction potential is the same at every point in the fluid, and thus reduces to a constant K. By identifying the two expressions of energy, he obtains the expression of c_α to within one multiplicative coefficient. He used Schiff's table of values for around forty organic compounds based on carbon (C), hydrogen (H), oxygen (O), and halogens such as chlorine (Cl), bromine (Br) and iodine (I) to determine the value of c_α for each species.

A remarkable point, noted by Boris Iglewicz[195], is Albert's use of the least squares technique to determine the value of the c_α that enter as linear combinations into the different molecules. The author compares the results of Albert's calculations, which were laborious for those who had to do them by hand, with the results that can be obtained today using a computer. Although there are disparities, sometimes significant, the agreement is generally satisfactory. Iglewicz writes in the summary of his article:

> His paper uses modelling and least squares to analyse data in support of a scientific proposition. Einstein is shown to be well trained, for his day, in using statistics as a tool in his scientific research.

The author believes that he became familiar with this technique by studying it himself. However, it should be remembered that in the second semester of 1897-1898, Albert attended Johann Jakob Rebstein's optional courses at ETH on the mathematical foundations of statistics and personal insurance, as we saw in the previous chapter. Rebstein was a remarkable calculator and a recognised specialist in the least squares technique, which he used in geodesics for the Swiss land register.

Albert's first article received little attention. In retrospect, however, its construction is very interesting. It shows him comparing the results of two different areas of physics to extract information about atoms, quantities whose physical reality had been called into question by some physicists, such as Ernst Mach and Georg Helm. Einstein's presentation was clear and logical, and his calculation techniques were well mastered. The letters to Mileva show that the determination of c_α was one of the objectives of his thesis. He still had the subject in mind at the end of 1901 when he wrote to her from Schaffhausen, after submitting his thesis manuscript to Kleiner:

> I got again a very self-evident but important scientific idea about molecular forces. You know that no noticeable evolution of heat takes place when two

---

[195] Boris Iglewicz, "Einstein's First Published Paper", *The American Statistician* **61** (4), 339-342 (2007).

neutral liquids are mixed together. From this it follows, according to our theory of molecular forces, that there must exist an approximate proportionality between our constants c_α and the molecular volumes of the liquids. If this were true, then this would be the end of the molecular-kinetic theory of liquids. I'll see whether I can get hold of Ostwald or Landolt during the vacation period. I'll either stay here (for reasons of economy) or go to Zurich and work (this beats all secondary considerations)[196].

This article on capillary phenomena was certainly a milestone in the thesis on molecular forces in liquids that he had begun in October 1900 after leaving ETH, since on December 20, 1900 Mileva wrote to her friend Helene Savić: "Albert is still here [in Zurich] and is going to stay here until he finishes his doctoral thesis, which will probably take until Easter". We shall see that this date was postponed by Albert because he suddenly saw a possible extension of his work to gases in April 1901.

Extending the subject of the thesis to molecular forces in gases

Albert did not have a job between August 1900 and May 1901, so he spent most of his time in Milan in March and April working on his bibliography. There he continued his work on molecular forces, which he planned to extend to gases. Mileva informs us of this in retrospect in a letter to Helene Savić in mid-December 1901:

> Albert has written a magnificent study, which he submitted as his dissertation. He will probably get his doctorate in a few months [...] I'll send you a copy when it gets printed. It deals with the investigation of the molecular forces in gases using various known phenomena[197].

This extension can be dated precisely to mid-April 1901. Albert had already spoken to his friend Marcel Grossmann about it on April 14:

> As for science, I have a few splendid ideas, which now only need proper incubation. I am now convinced that my theory of atomic attraction forces can also be extended to gases, and that it will be possible to obtain the characteristic constants of almost all elements without great difficulty. That will then also bring the problem of the inner kinship between molecular forces and Newtonian action-at-a-distance forces much nearer to its solution. It is possible that experiments already done by others for other purposes will suffice for the testing of the theory. In that case I shall utilize all the already existing results in my doctoral dissertation. It is a

---

[196] Letter to Mileva Marić in *CPAE (English translation supplement)*, vol. 1, Doc. 127, pp. 185-186.
[197] Letter from Mileva Marić to Helene Savić in *CPAE (English translation supplement)*, vol. 1, Doc. 125, p. 183.

glorious feeling to perceive the unity of a complex of phenomena which appear as completely separate entities to direct sensory observation[198].

He confirms this point of view the next day by addressing Mileva in similar terms:

> As for science, I've got an extremely lucky idea, which will make it possible to apply our theory of molecular forces to gases as well. You certainly remember that the force function appears explicitly in the integrals that have to be evaluated for the calculation of diffusion, thermal conduction & viscosity. Hence, with gas molecules, *only* our constants c_α are necessary for the calculation of these coefficients for ideal gases, and one does not have to venture into the theoretically so uncertain area of deviations from the ideal gas state. I can hardly await the outcome of this investigation. If it leads to something, we will know almost as much about the molecular forces as about the gravitational forces, and only the law of the radius will still remain unknown. Unfortunately, I must also admit that this idea for the investigation of salt solutions rested on such a weak basis that I think that one should first restrict oneself to the investigation of infinitely dilute solutions, in which an interaction between the molecules of the dissolved substance does not yet occur. One can so determine a great number of c_α's, which could be used for an approximate verification of the hypothesis of the kinship with gravitation. It is possible that information about the law of action itself will more likely be provided by the quantities $(\gamma - Td\gamma/dT)$ / volume and the integrals from the theory of gases[199].

It should be noted that these two letters of mid-April were sent from Milan, where he had been staying since March 23. So it was in Milan that he had the idea for this opening for his thesis, a city where we know he worked in the library of the Lombard Institute. Could reading an article in this library have motivated him?

Max Reinganum's article in the Lorentz Jubilee volume

Among the topical events of the day, the first was the *Jubilee* for the Dutch physicist Hendrik Antoon Lorentz, held in Leiden on December 11, 1900 to mark the twenty-fifth anniversary of his doctoral thesis. The organisers had asked leading physicists to send in their contributions a year in advance. The *Jubilee* was published in a special volume of almost seven hundred pages,

---

[198] Letter to Marcel Grossmann in *CPAE (English translation supplement)*, vol. 1, Doc. 100, p. 165.
[199] Letter to Mileva Marić in *CPAE (English translation supplement)*, vol. 1, Doc. 101, p. 166.

Volume V of the *Dutch Archives of Natural Sciences*[200]. This volume was received by the library of the Lombard Institute on January 31, 1901, as can be seen from the hand-inscribed mark on the cover of the library copy. It was therefore available there when Albert was in Milan in the spring of 1901. The volume contains texts in German, French, English and Italian. The subjects cover most fields of physics, and several articles are likely to have been of interest to him, such as those on electrodynamics and radiation by Emil Cohn, Julius Farkas, Dimitrij Goldhammer, Walter Kaufmann, Max Planck, Henri Poincaré, Augusto Righi, Emil Wiechert and Wilhelm Wien.

In the field of molecular forces, the *Jubilee* contains Max Reinganum's paper on "Molecular attraction in weakly compressed gases", the very title of which suggests that it certainly played a role in expanding Albert's subject. Reinganum sought to determine the equation of state of gases on the basis of "planetary" forces of attraction, similar to gravitation but differing from it in their dependence on distance. According to him, the size of molecules plays no role in dissociation processes, so he considers them to be centres of force. This point of view, which contradicted that of Boltzmann and van der Waals, was nevertheless the one adopted by Albert, who reminded Mileva of this at the end of April:

> I am very curious whether our conservative molecular forces will hold good for gases as well. If here too the mathematically so unclear concept of molecular size does not manifest itself in the formation of the trajectories of molecules coming close to each other, but the molecule can be conceived as centre of force. We shall get quite a precise test of our view[201].

Albert could therefore find support for his ideas in Reinganum. He knew this author, who was quoted by Boltzmann in his lectures. Perhaps even the letter of application for an assistant's post that he sent on April 12 to Kamerlingh Onnes in Leiden, the patron of Reinganum, mentioned in the previous chapter, was not independent of his reading of the article[202]? It should be noted that the following month Albert was interested in another article by Reinganum on Drude's theory of electrons. April 1901 was also the period when he discussed the dissociation of molecules with Michele, probably in connection with his reading of the chapter VI entitled "Theory of dissociation" in Boltzmann's lectures on gas theory, Part II. The author admits that he could not do without the size of the molecules introduced by van der Waals, which was essential to his theory. It cannot be ruled out that

---

[200] *Archives néerlandaises des sciences exactes et naturelles*, vol. 5, series 2 (1900): the *Jubilee*.

[201] Letter to Mileva Marić in *CPAE (English translation supplement)*, vol. 1, Doc. 102, pp. 167-168.

[202] The *CPAE*, vol. 1, note 6, Doc. 111, p. 305, refer to Reinganum Max, "Theoretische Bestimmung des Verhältnisses von Wärme - und Elektricitätsleitung der Metalle aus der Drude'schen Elektronentheorie", *Annalen der Physik* **2**, 398-403 (1900).

Boltzmann's last sentence, which states that "A mechanical model based only on attractive forces, without elastic repulsive forces in collisions, agreeing with all experimental facts for the gaseous and liquid states of aggregation, has not yet been found", appeared to Albert as a challenge and a good reason to turn to Reinganum.

Abandoning the thesis in February 1902

Albert wrote to Mileva from Schaffhausen on November 28, 1901, in a tone that hinted at some doubts for the future: "So far I have no report from Kleiner. I don't think he would dare to reject my dissertation, but otherwise, in my opinion, there is nothing that can be done with that short-sighted man[203]". With no news of his manuscript, he then undertook to chase after Kleiner and recounted two days later his meeting to Mileva:

> Today I spent the whole afternoon with Kleiner in Zurich and explained my ideas on the electrodynamics of moving bodies to him & otherwise talked with him about all kinds of physical problems. He is not quite as stupid as I thought, and, moreover, he is a good guy. He said I may refer to him whenever I need a recommendation. Isn't that nice of him? He must go away during the vacation and he hasn't read the thesis yet[204].

It seems that the subject of molecular forces, so dear to Albert, was not discussed, Kleiner probably waiting for an expert opinion before making a decision. A few days later, on December 28, Albert's hopes for his thesis were at their highest, as he wrote victoriously to Mileva that, compared with his friends Jakob Ehrat and Marcel Grossmann, he would be "the first to finish his dissertation". However, on February 1, 1902, he asked to be reimbursed for his tuition fees at the University of Zurich, which he had paid on November 23. Albert did not submit his thesis, again with Kleiner, until three years later, in 1905, with a title that contrasted with his earlier work: *A new determination of molecular dimensions.*

The reasons for abandoning the 1901 thesis are not known with certainty. According to Rudolf Kayser, Albert's son-in-law, "Kleiner had rejected this essay out of consideration for his colleague Ludwig Boltzmann, whose train of reasoning Einstein had sharply criticized"[205]. In February 1902, Albert told Mileva about the advice of his friend Habicht: "I am now expounding to Habicht the paper I have handed in to Kleiner. He is quite enthusiastic about

---

[203] Letter to Mileva Marić in *CPAE (English translation supplement)*, vol. 1, Doc. 126, p. 184.

[204] Letter to Mileva Marić in *CPAE (English translation supplement)*, vol. 1, Doc. 130, pp. 188-189.

[205] Anton Reiser, *Albert Einstein: A Biographical Portrait*, London, Thornton Butterworth, 1931, p. 71.

my good ideas & nags me to send Boltzmann the part of the paper that refers to his book. I am going to do it[206]". However, the publication that followed in 1902 in the *Annalen der Physik,* entitled "Kinetic theory of thermal equilibrium and the second law of thermodynamics", had no direct connection with molecular forces, but was linked to his further reading of Boltzmann, in particular the theoretical part of chapters that he had probably skimmed over previously[207]. However, it is highly likely that if Kleiner had asked Boltzmann for his opinion, the latter would have expressed his disagreement with a theory that did not take account of the size of molecules, a notion accepted by almost all specialists in the field since van der Waals.

It should be noted that later, in April 1911, Albert explicitly referred to Reinganum in a letter to his thesis student Hans Tanner, in connection with a work on viscosity published in 1901, but immediately distanced himself: "If it will not [suffice], then we will add something about viscosity, which, to be sure, will not be definitive, but will still be more consistent than Reinganum's work, because, for example, we count only collisions of nuclei as collisions[208]". A year later, he wrote to Kleiner about candidates to succeed Peter Debye at the University of Zurich: "From among the men you mentioned, I would first rule out Reinganum, Born, and Happel", because "Reinganum's papers are rather sloppy[209]". The extension of the subject of molecular forces to gases on the basis of a paper by Reinganum in mid-April 1901 was certainly also one of the "bad leads" mentioned by Michele.

Questions about the nature of light and light quanta from 1901?

The problem of the nature of light had preoccupied Albert since his student days and he was probably the only physicist to address it openly in 1900. As we have seen, in 1945 Michele Besso dated their first discussion on the nature of light "between Newton and Huygens", *i.e.*, between particles and waves, to around 1895. It was not until March 1899 that Albert informed

---

[206] Letter to Mileva Marić in *CPAE (English translation supplement)*, vol. 1, Doc. 136, p. 192.

[207] For a discussion, see C. Bracco & J.-P. Provost, Vers une meilleure compréhension des premières idées scientifiques d'Albert Einstein – son environnement scientifique en Italie (1895-1902), *Rendiconti (Scienze) Istituto lombardo* **152**, 113-154 (2018).

[208] Letter to Hans Tanner in *CPAE (English translation supplement)*, vol. 5, Doc. 265, p. 186. The editors of volume 5 refer to Max Reinganum's article "Über die Theorie der Zustandsgleichung und der inneren Reibung der Gase", *Physikalische Zeitschrift* **2**, 241-245 (1901).

[209] Letter to Alfred Kleiner in *CPAE (English translation supplement)*, vol. 5, Doc. 381, p. 285.

Mileva that his "broodings about radiation are starting to get on somewhat firmer ground and I am curious myself whether something will come out of them[210]", without going into further detail. Two years later, in the spring of 1901 in Milan, he became interested in the interaction of radiation with matter, and the reflections quickly followed one another. On March 23, after reading an article by Max Planck, he wrote to her, as mentioned in the previous chapter, that it seemed to him "that the latent kinetic energy of heat in solids and liquids can be conceived of as the energy of electrical resonators". On April 4, he informed Mileva that: "About Max Planck's studies on radiation, misgivings of a fundamental nature have arisen in my mind, so that I am reading his article with mixed feelings". On April 10, he explained his doubts by referring to the very basis of Planck's analysis, which "assumes that a completely definite kind of resonators (fixed period and damping) causes the conversion of energy to radiation, an assumption I cannot really warm up to". It should be noted in passing that this reference to well-defined damping of oscillators can only refer to Planck's article in Lorentz *Jubilee*. In fact, in his earlier articles, Planck made the mass and damping of resonators tend towards zero[211]. By this time, his "views on the nature of radiation have again sunk back into the sea of haziness". On leaving for Winterthur, he wrote to his girlfriend on April 30, 1901, still from Milan:

> It occurred to me recently that when light is generated, direct conversion of motional energy to light may take place because of the parallelism kinetic energy of the molecules – absolute temperature – spectrum (radiating space energy in the state of equilibrium). Who knows when a tunnel will be dug through these hard mountains![212]

Albert uses a railway metaphor! He is referring to Wien's work of 1896[213]. Wien had obtained his spectral law of blackbody radiation by using Maxwell's law of velocity distribution for the gas molecules contained in the enclosure and then introducing an *ad hoc* relationship of proportionality between the kinetic energy of these molecules and the frequency of the radiation they generate in the ether. Albert was familiar with Wien's law, if only through the electrical journals he had been able to consult in his uncle Jakob's office in Pavia from 1895, in connection with the lighting of incandescent lamps. The following month, he read Philipp Lenard's experimental

---

[210] Letter to Mileva Marić in *CPAE (English translation supplement)*, vol. 1, Doc. 45, p. 126.

[211] Comment by Jean-Pierre Provost.

[212] Letter to Mileva Marić in *CPAE (English translation supplement)*, vol. 1, Doc. 102, pp. 167-168.

[213] Wien Willy, "Ueber die Energievertheilung im Emissionsspectrum eines schwarzen Körpers", *Annalen der Physik* **294** (8), 662-669 (1896).

article on the photoelectric effect[214], which filled him with joy, and wrote to Mileva on May 28:

> I have just read a marvellous paper by Lenard on the production of cathode rays by ultraviolet light. Under the influence of this beautiful piece of work, I am filled with such happiness and such joy that you absolutely must share in some of it[215].

In this article, Lenard unambiguously states that "cathode rays" (Joseph John Thomson's electrons) originate from the metal cathode illuminated by ultraviolet light.

How else can we understand Albert's elation than by thinking that he now realised that the photoelectric effect corresponds to the opposite process to that at work in emission with Wien's law? The kinetic energy transferred by the molecules to the radiation thus appears to be convertible by absorption of the radiation into the kinetic energy of the electrons, causing them to be ejected from the metal! Planck's quanta can therefore provide a simple explanation of the process if they are given their own reality as light particles. In his interviews with Robert Shankland in 1952, Albert pointed out that "the photoelectric effect paper was also the result of five years pondering and attempts to explain Planck's quantum in more specific terms". As Albert only became aware of Planck's paper in April 1901 and at that time began to rethink the nature of radiation, it is understandable that he wrote to Michele Besso in 1951: "A total of fifty years of conscious speculation has not brought me any closer to answering the question 'What are light quanta?'". The idea of light quanta as early as 1901 was first suggested by Jürgen Renn in 1993, essentially on the basis of the arguments recalled above and Albert's atomistic view of matter[216].

It is also possible that a result contained in Poincaré's article entitled "Lorentz's theory and the principle of reaction", which also appears in Lorentz *Jubilee*, an article that Albert cited in 1906 in his analysis of the motion of the centre of mass, contributed to his kinetic view of light. In this article, Poincaré justifies, in order to satisfy the law of action and reaction in Lorentz's theory, that the electromagnetic field must carry a momentum. For a piece of plane wave, this momentum is linked to energy by the relation $p = E/c$. Using the Lorentz transformations of 1895, Poincaré calculated the length, energy and momentum (energy divided by c) of a piece of plane wave emitted by a Hertzian oscillator located at the focus of a parabolic mirror, in

---

[214] Lenard Philipp, "Erzeugung von Kathodenstrahlen durch ultraviolettes Licht", *Annalen der Physik* **2**, 359-375 (1900).
[215] Letter to Mileva Marić in *CPAE (English translation supplement)*, vol. 1, Doc. 111, p. 174.
[216] Jürgen Renn, "Einstein as a Disciple of Galileo: A Comparative Study of Concept Development in Physics", *Science in Context* **6** (1), 311-341 (1993).

order to study the recoil of the mirror. He first carried out the calculation in a frame of reference at rest in the ether, then in a frame of reference moving at speed V relative to the ether. He shows that the length of the wave train (*i.e.* the wavelength) is transformed as $1 + V/c$ and its energy as $1 - V/c$. Now, for a wave train at the frequency v, the relation $\lambda = c/v$, which relates it to the wavelength, indicates that the frequency is transformed like the energy, *i.e.* into $1 - V/c$. For this wave train, this suggests a proportionality between energy and frequency[217].

It should be pointed out that neither Planck nor Poincaré, who presented their results four days apart in December 1900, anticipated the existence of light quanta. Planck remained faithful to a classical, Maxwellian, continuous vision of radiation, and sought to focus the discretisation on energy exchanges with matter. Poincaré did not take a position on the quantum hypothesis until 1911, when he was invited to the Solvay Congress. He did so to prove mathematically the inescapable nature of the discontinuity hypothesis introduced by Planck and published in 1912 three papers in a row.

Albert could thus have conceived, on the basis of a synthesis of the various works mentioned above, as early as 1901, a naive model of light quanta, comparable to a piece of plane wave of volume λ^3 and energy hv[218], as he gave the image in his Salzburg lecture to the German Physical Society in 1909[219]. But to further justify this relationship in the eyes of physicists, he would have to wait until he had an argument based on statistical physics, on which he was working between 1902 and 1904. In the conclusion of an article devoted to fluctuations ΔE in the energy E of black-body radiation, entitled "On the general molecular theory of heat[220]" and sent to the *Annalen der Physik* on March 27, 1904, he showed that by applying the fluctuations formula to the equality $\Delta E = E$ for radiation of volume $V = \lambda^3$, *i.e.* for a

---

[217] Christian Bracco and Jean-Pierre Provost, "La relativité de Poincaré de 1905", in Tahar Boudjedaa, Abdenacer Makhlouf (eds.), *Théorie quantique des champs : Méthodes et applications*, Actes de l'école de physique théorique de Jijel, 2006, Paris, Hermann, coll. "Travaux en cours", vol. 68, 2007, pp. 323-354; Christian Bracco, "L'environnement scientifique du jeune Albert Einstein : la période milanaise 1899-1901", *Revue d'histoire des sciences* **68** (1), 109-144 (2015).

[218] Indeed, Stefan's law $E \propto VT^4$ giving the energy of radiation in a volume V at temperature T is also written $E \propto v$ if applied to a packet of light of volume $V = \lambda^3$ and taking into account Wien's law λT or T/v = constant: note 88 by Jean-Pierre Provost, in C. Bracco, "L'environnement scientifique du jeune Albert Einstein : la période milanaise 1899-1901", art. cit.

[219] "On the development of our views concerning the nature and constitution of radiation" [*Deutsche Physikalische Gesellschaft, Verhandlungen* **7**, 482-500 (1909)] in *CPAE (English translation supplement)*, vol. 2, Doc. 60, pp. 379-394.

[220] "On the general molecular theory of heat" [*Annalen der Physik* **14**, 354-362 (1904)] in *CPAE (English translation supplement)*, vol. 2, Doc. 5, pp. 68-77.

quanta, he indeed found Wien's law of displacement[221]. Probably the arrival of his friend Michele Besso at the Patent Office in Berne in January 1904, when he was writing this article, and a return to their earlier discussions in Milan in April 1901, were not unrelated to this consideration? This would enable us to understand Michele's repeated references to 1904 on the subject of quanta, recalled in the introduction, and his mysterious reference to "Planck's quanta, which were still very young and had already been forgotten in 1904". However, it was not until 1905, on the basis of analogical reasoning based on the volume dependence of the entropy of radiation, still within the limits of Wien's law of 1896[222], that Albert was finally able to demonstrate the existence of independent light quanta, a truly "revolutionary" hypothesis.

---

[221] Robert Rynasiewicz and Jürgen Renn, "The Turning Point for Einstein's *Annus Mirabilis*", *Studies in History and Philosophy of Modern Physics* **37**, 5-35 (2006).

[222] "On a heuristic point of view concerning the production and transformation of light" [*Annalen der Physik* **17**, 132-148 (1905)] in *CPAE (English translation supplement)*, vol. 2, Doc. 5, pp. 86-103.

A final word

By way of conclusion to our discussion of the young Albert's Italian period, in Pavia and Milan, let us point out that the Lombard scientific community did not forget him. In May 1922, Giuseppe Jung, Michele's uncle who provided him with connections in Milan, was one of the five nominators, and in this case the first on the list, along with Luigi Berzolari, Antonio Federico Jorini, Francesco Gerbaldi and Giulio Vivanti, for his appointment as a foreign member of the Lombard Institute, alongside Gösta Mittag-Leffler and Paul Painlevé, and three Italian members. The proposal was drafted on May 4, 1922 by Giulio Vivanti, whom we had met at the University of Pavia, where he was teaching analysis and algebra at the time when Albert was preparing for the ETH in Zurich. The election took place in July 1922, a few months before the Nobel Prize was awarded. Einstein and Mittag-Leffler were elected with fourteen out of sixteen votes, the others unanimously. The text of the proposal highlighted Albert Einstein's work on special relativity and general relativity, in a very explicit summary:

> Albert Einstein's name needs no comment. Although the theory of relativity, like all the others, had its precursors, there can be no doubt that Einstein deserves credit for having boldly stated the fundamental postulates of the independence of physical phenomena from the uniform rectilinear translation of a reference system, and of the absolute constancy of the speed of light in a vacuum. Later, as the theory on which he had laid the foundations was developing strongly, the need to extend its field of application led him to modify his postulates by asserting the invariance of physical phenomena with respect to changes in more general reference systems. The discussions that the new theory has provoked in the scientific world, the keen curiosity with which experimental confirmations are awaited, the echo that it has met even outside the circle of scientists – all of this shows that this exceptional sum is universally attributed to Einstein's views, so much so that even those who disagree with him on any particular point recognise him as a renovator of natural philosophy.

Albert was already enjoying widespread popularity at the time, following the verification of an essential prediction of general relativity by the astronomer Arthur Eddington in 1919: the deviation of starlight as it passes close to the Sun, observable during an eclipse. In April 1922, Albert Einstein had just given a series of lectures on general relativity at the Collège de France in Paris, at the invitation of Paul Painlevé. Albert's old connections in Pavia and Milan had thus been transformed into support for his Nobel Prize in 1922.

Index of names

Ansbacher, Bernardo 13, 96, 110, 111
Ansbacher, Luigi 107, 111
Aron, Hermann 46, 51
Ascoli, Graziadio 24, 30, 32, 64, 68
Ascoli, Moisè 30, 31, 66, 68, 103

Barberis, Giovanni 16, 32, 59, 62, 112
Battelli, Angelo 73, 101, 108, 110
Besso, Beniamino 8, 12, 25, 55, 56, 62, 63, 64, 65, 66
Besso, Davide 12, 55, 63, 64, 68, 69, 110
Besso, Giuseppe 12, 55, 59, 61, 63, 64, 71, 72
Besso, Marco 7, 12, 24, 27, 55, 59, 62, 63, 64, 66, 67, 68, 69
Besso, Michele 3, 4, 5, 6, 7, 8, 9, 10, 12, 36, 55, 56, 57, 58, 59, 60, 61, 62, 63, 71, 73, 74, 83, 95, 96, 97, 99, 100, 101, 102, 103, 104, 110, 112, 113, 114, 115, 118, 123, 125, 127, 129
Blaserna, Pietro 23, 30, 31, 44, 56, 65, 68, 73, 103
Boltzmann, Ludwig 8, 98, 104, 106, 108, 120, 123, 124
Brioschi, Francesco 18, 19, 23, 30, 32, 73, 101

Cantoni, Angelo 61, 70
Cantoni, Erminia 61, 64, 70
Cantoni, Eugenio 23, 61, 70
Cantoni, Giovanni 22, 32, 43, 44, 68, 76, 85, 87, 88, 91, 103
Cantoni, Vittorio 12, 30, 49, 55, 62, 70, 71, 140
Cattaneo, Carlo 21, 22, 30, 87
Cerri, Angelo 12, 16, 84, 85, 90

Colombo, Giuseppe 11, 17, 18, 19, 24, 26, 28, 30, 31, 41, 44, 47, 49, 97, 102, 110

Drude, Paul 83, 97, 105, 106, 108

Edison, Thomas 26, 27, 35, 37, 41, 45, 49, 51
Einstein, Hermann 10, 11, 16, 35, 83, 106
Einstein, Jakob 10, 11, 15, 16, 35, 45, 47, 49, 51, 54, 62, 83, 86
Einstein, Maja 4, 12, 15, 18, 19, 57, 62, 77, 82, 84, 89, 106, 111
Einstein, Rudolph 61, 107

Ferraris, Galileo 11, 29, 31, 36, 44, 46, 47, 49, 50, 51, 62, 65, 71, 75, 77
Ferrini, Rinaldo 26, 29, 49

Gabba, Luigi 13, 96, 109, 110, 111, 112
Garrone, Lorenzo 12, 16, 17, 27, 31, 35, 50, 52, 83
Gaulard, Lucien 11, 36, 48, 49, 53
Gibbs, John Dixon 11, 36, 48, 49
Gibbs, Willard 119
Grossmann, Marcel 4, 7, 109, 112, 116, 121

Heine, Eduard 104, 118
Helmholtz, Hermann von 36, 43, 50, 53, 98
Hertz, Heinrich 81, 99, 100
Herzog, Albin 57, 74, 75, 78, 81
Hoepli, Ulrico 32, 33, 112

Jung, Giuseppe 12, 13, 30, 55, 59, 73, 74, 96, 100, 101, 108, 109, 131

Kayser, Rudolf 1, 25, 98, 124
Kleiner, Alfred 116, 120, 124, 125
Koch, Caesar 80
Koch, Jokob 112
Koch, Pauline 15

Larmor, Joseph 81, 82, 94, 102
Lenard, Philipp 126
Liénard, Alfred 53, 117
Lodge, Oliver 82, 100
Lorentz, Hendrik Antoon 8, 53, 81,
    102, 117, 122, 127
Luzzatti, Luigi 19, 27, 30, 32, 62, 68

Mach, Ernst 6, 36, 43, 46, 57, 98, 114,
    115, 120
Maffei, Clara 18, 72
Maier, Gustav 61, 75, 78, 81
Marangoni, Carlo 12, 76, 79, 81, 88,
    89, 90, 91, 92, 119
Marangoni, Ernestina 12, 18, 81, 84, 89
Marić, Mileva 2, 3, 5, 59, 60, 68, 73,
    95, 96, 97, 98, 99, 100, 103, 104,
    105, 106, 107, 108, 109, 112, 114,
    115, 116, 117, 120, 121, 122, 123,
    124, 125, 126, 127
Maxwell, James Clerk 53, 99, 101
Melloni, Macedonio 118
Miller, Oskar von 44, 45, 86
Minkowski, Hermann 92, 119
Monti, Carlo 17, 26, 112

Neustätter, Otto 76, 86, 89
Nobili, Leopoldo 118

Planck, Max 2, 6, 8, 13, 97, 105, 106,
    123, 127, 128
Poincaré, Henri 53, 69, 83, 100, 102,
    117, 119, 123, 127, 128

Rebstein, Jakob Johann 108, 120
Reinganum, Max 13, 105, 108, 113,
    123, 124, 125
Richard, Gustave 40, 53
Righi, Augusto 56, 73, 100, 101, 103,
    108, 110, 123
Rossetti, Francesco 39, 43, 88

Schiff, Roberto 69, 92, 119
Seelig, Carl 1, 5, 6, 7, 10, 15, 61, 77,
    104
Shankland, Robert 2, 88, 127
Speziali, Pierre 5, 19, 55, 56, 57, 58,
    60, 61, 63, 64, 70

Thomson, William (Lord Kelvin) 33,
    37, 43, 93, 94, 119

Violle, Jules 33, 36, 39, 40, 82, 92
Vivanti, Giulio 84, 131
Volta, Alessandro 17, 23, 31, 43, 98

Weber, Heinrich Friedrich 11, 36,
    44, 50, 57, 72, 74, 75, 76, 77,
    104, 107
Wien, Wilhelm 96, 115, 123, 126
Winteler, Anna 57, 58
Winteler, Jost 19, 57, 61, 81, 108

Selected bibliography

Bernini Fabrizio, *Che bel ricordo Casteggio... : Albert e Maja Einstein ed il salotto letterario di Ernestina Pelizza Marangoni*, Edizioni Modulo Tre di Diego Dabusti & C., 1994.

Besso Marco, *Autobiografia*, Rome, Fondazione Marco Besso Editrice, 1925, repr. 1970.

Bevilacqua Fabio and Renn Jürgen (Eds.), *Albert Einstein, ingegnere dell'universo*, Milan, Skira, 2005.

Biscossa Sergio, *Gli Einstein imprenditori a Pavia (1894-1896)*, Mortara, Rotary Club di Vigevano, 2005.

Boltzmann Ludwig, *Vorlesungen über Gastheorie*, Leipzig, J.A. Barth, 1896-1898; translated by Stephen G. Brush, *Lectures on gas theory*, London, Cambridge University Press, 1964.

Borghi Emilio, "L'attività scientifica e didattica di Carlo Marangoni nel R. Liceo Dante di Firenze", *Atti della Fondazione Giorgio Ronchi* **3**, 287-332 (2008).

Bosscha Johannes (Ed.), *Jubilee for Lorentz (Festschrift)*, *Archives néerlandaises des sciences exactes et naturelles* **5**, serie 2, La Haye, Martinus Nijhoff, 1900.

Bracco Christian, "Einstein and Besso: From Zurich to Milan", *Istituto Lombardo (Rend. Scienze)* **148**, 258-322 (2014); http://www.ilasl.org/index.php/Scienze/article/view/178

Bracco Christian, "L'environnement scientifique du jeune Albert Einstein : la période milanaise 1899-1901", *Revue d'histoire des sciences* **68** (1), 109-144 (2015).

Bracco Christian, "An overview of Albert Einstein's and Michele Besso's links with Italian universities and engineering schools", *Annali di storia delle università italiane* **19** (2), 129-152 (2015).

Bracco Christian and Provost Jean-Pierre, "Perspective on Einstein's scientific work in Milan", *Proccedings of the 14th Marcel Grossmann meeting on General Relativity*, 12-18 July 2015, Massimo Bianchi, Robert Jantzen and Remo Ruffini (Eds.), Singapore, World Scientific, 2018, pp. 3337-3341.

Bracco Christian and Provost Jean-Pierre, Vers une meilleure compréhension des premières idées scientifiques d'Albert Einstein – son environnement scientifique en Italie (1895-1902), *Rendiconti (Scienze) Istituto lombardo* **152**, 113-154 (2018).

Caracciolo Alberto, "Una Diaspora di Trieste: I Besso nell'Ottocento", *Quaderni storici* **18** (54/3) Ebrei in Italia, 897-912 (1983).

Casella Antonio and Lucchini Guido, *Graziadio e Moisè Ascoli. Scienza, cultura e politica nell'Italia liberale*, Pavia, La Goliardica Pavese, 2002.

Cattaneo Carlo, *The Milan Uprising*, in *Tutte le Opere di Carlo Cattaneo*, IV, Luigi Ambrosoli (Ed.), Verona, Mondadori, 1967.

Darrigol Olivier, *Electrodynamics from Ampère to Einstein*, Oxford, Oxford University Press, 2000.

Darrigol Olivier, "Poincaré, Einstein et l'inertie de l'énergie", *Comptes rendus de l'Académie des sciences* **1** (1), 143-153 (2000).

Décaillot Anne-Marie, *Cantor et la France. Correspondance du mathématicien allemand avec les Français à la fin du XIXe siècle*, Paris, Kimé, 2008.

Decleva Enrico, *Ulrico Hoepli (1847-1935) editore e libraio*, Milan, Hoepli, 2001.

Eine neue Zeit! Die internationale elektrotechnische Ausstellung 1891, Frankfurt, Frankfurt am Main Historisches Museum, 1991.

Einstein Albert, "Notice for an autobiography", *The Saturday review of literature* **26**, 9-13 (1949).

Einstein Albert, "Autobiographische Skizze", in Carl Seelig, *Helle Zeit – Dunkle Zeit*, Zurich, Europa Verlag, 1956, pp. 9-17.

Einstein Albert, "How I Created the Theory of Relativity", translated by Yoshimasa Ono from notes in Japanese by J. Ishiwara, *Physics Today* **35** (8), 45-47 (1982).

Exposition internationale d'électricité Paris 1881, Administration-Jury-Rapports, t. 1, 2 vols, Paris, Masson, 1883.

Fölsing Albrecht, *Albert Einstein: a biography*, New York, Viking, 1997 (Penguin Books, 1998). English translation from German.

Frank Philpp, *Einstein: his life and times*, London, Jonathan Cape, 1949.

Fregonese Lucio, *Gioventù felice in terra pavese. Le lettere di Albert Einstein al Museo per la Storia dell'Università di Pavia*, Milan, Cisalpino Istituto Editoriale Universitario, 2005.

Gatti Emilio and Robbiati Bianchi Adele (Eds.), *Istituto lombardo accademia di scienze e lettere*, 3 vols, Milan, Libri Scheiwiller, 2007-2009.

Gobbi Ulisse, *Le societa di mutuo soccorso*, Milan, Società editrice libraria, 1901.

Gobbo Raffaella, "L'archivio di Galileo Ferraris", *Rassegna degli archivi di stato, nuova serie* **1** (1-2), 9-169 (2005).

Gobbo Raffaella and Silvestri Andrea (Eds.), *L'archivio di Galileo Ferraris*, vol. 1, *Corrispondenza-Inventario*, Vercelli, Gallo, 1997.

Guareschi Icilio, "Nota sulla storia del movimento browniano", *Isis* **1**, 47-52 (1913).

Heine Eduard, *Handbuch der Kugelfunctionen, Theorie und Anwendungen*, 2 vols, Berlin, Reimer, 1878 (vol. 1) and 1881 (vol. 2).

Hettler Nicolaus, *Die Elektrotechnische Firma J. Einstein & Cie in München – 1876-1894*, Universität Stuttgart, Grin e-Book, 1996.

Iglewicz Boris, "Einstein's First Published Paper", *The American Statistician* **61** (4), 339-342 (2007).

Katzir Shaul, "Hermann Aron's Electricity Meters: Physics and Innovation in Late Nineteenth-century Germany", *Historical Studies in the Natural Sciences* **39** (4), 444-481 (2009).

Lodari Renata, *Giardini e ville del lago Maggiore. Un paesaggio culturale tra Ottocento e Novecento*, Turin, Centro Studi Piemontesi, 2002.

Lori Ferdinando, *Storia del R. Politecnico di Milano*, Milan, Cordani, 1941.

Maggi Gian Antonio, "Giuseppe Jung, Commemorazione letta nell'adunanza del 14 aprile 1927", *Rendiconti Istituto lombardo* **60** (2), 291-307 (1927).

Maifreda Germano, *Gli Ebrei e l'economia milanese: l'Ottocento*, Milan, FrancoAngeli, 2000.

Marchis Vittorio (Ed.), *Letture Politecniche I (1889-1906)*, Turin, Centro Studi Piemontesi, 2008.

Marcillac P., "Le transport d'énergie électrique de Tivoli à Rome", *La Lumière électrique* **40**, 7-16 (1893).

Marcolongo Roberto, "Davide Besso", *Periodico di Matematica* **22**, 147-156 (1907).

Milza Pierre, *Histoire de l'Italie, des origines à nos jours*, Fayard/Pluriel, 2013.

Minkowski Hermann, "Kapillarität", in *Enzyklopädie der mathematischen Wissenschaften mit Einschluss ihrer Anwendungen*, Bd. 5, *Physik*, edited Arnold Sommerfeld, Leipzig, Teubner, 1903-1926, Teil 1, pp. 558-613.

Mori Giorgio (Ed.), *Storia dell'industria elettrica in Italia*, vol. 1, *Le origini, 1882-1914*, Bari, Laterza, 1992.

Pais Abraham, *Subtle is the Lord: the science and the life of Albert Einstein*, Oxford, Oxford University Press, 1982.

Parville Henri de, *L'Électricité et ses applications – Exposition de Paris*, Paris, Masson, 1882.

Poincaré Henri, *Électricité et optique*, Paris, Carré & Naud, 2nd edn, 1901.

Poincaré Henri, "La théorie de Lorentz et le principe de réaction", *Archives Néerlandaises des Sciences Exactes et Naturelles* **5**, 252-278 (1900).

Pyenson Lewis, *The young Einstein: the advent of relativity*, Bristol, Adam Hilger, 1985.

Reinganum Max, "Molekuläre Anziehung in schwach comprimierten Gasen", *Archives Néerlandaises des Sciences Exactes et Naturelles* **5**, 574-582 (1900).

Reiser Anton (pseudonym of Rudolph Kayser), *Albert Einstein: A Biographical Portrait*, London, Thornton Butterworth, 1931.

Renn Jürgen, "Einstein's controversy with Drude and the origin of statistical mechanics: A new glimpse from the 'Love Letters'", *Archive for the history of exact sciences* **51**, 315-354 (1997).

Renn Jürgen, "Enstein as a Disciple of Galileo: a Comparative Study of Concept Development in Physics", *Science in Context* **6** (1), 311-341 (1993).

Renn Jürgen and Schulmann Robert (Eds.), *Einstein Albert, Marić Mileva: The Love Letters*, Princeton, Princeton University Press, 1992.

Richard Gustave, "Les lampes à arc", *La Lumière électrique* **49**, 313-320 (1893).

Rynasiewicz Robert and Renn Jürgen, "The Turning Point for Enstein's Annus Mirabilis", *Studies in History and Philosophy of Modern Physics* **37**, 5-35 (2006).

Sacheri Giovanni (engineer) (Ed.), *L'ingegneria, le arti e le industrie alla esposizione generale italiana in Torino 1884*, Rivista tecnica compilata, Turin, 1890.

Schram Albert, *Railways and the Formation of the Italian State in the Nineteenth Century*, Cambridge, Cambridge University Press, 1997.

Seelig Carl, *Albert Einstein und die Schweiz*, Europa Verlag, Zurich-Suttgart-Vienna, 1952 translation London, Staples Press, 1956.

Shankland Robert, "Conversations with Albert Einstein", *American Journal of Physics* **31**, 47-57 (1963).

Silvestri Andrea (Ed.), *Francesco Brioschi (1824-1897)*, Convegno di studi matematici, 22-23 Ottobre 1997, Milan, Istituto lombardo di scienze e lettere, 1999.

Silvestri Andrea, *Omaggio ad Albert Einstein*, Milan, Politecnico di Milano, 2005.

Speziali Pierre (notes and introduction by), *Einstein Albert, Besso Michele, Correspondance (1903-1955)*, French edition, Paris, Hermann, 1979.

Stachel John, *Einstein's Miraculous Year: Five Papers That Changed the Face of Physics*, Princeton, Princeton University Press, 2005.

Thomson William (Sir), *Popular Lectures and Addresses*, vol. 1, London, Macmillan & Co, 1889.

Trbuhović-Gjurić Desanka, *Mileva Einstein, une vie*, translation by Nicole Casanova, Paris, Des Femmes, 1991.

Violle Jules, *Lehrbuch der Physik*, 2 vols, Berlin, Springer, 1892 (vol. 1) and 1893 (vol. 2).

Wertheimer Max, *Productive Thinking*, New York and London, Harper and Brothers, 1945.

The Collected Papers of Albert Einstein [CPAE]:
<https://einsteinpapers.press.princeton.edu/>

Stachel John, Cassidy David C., and Schulmann Robert (Eds.), *CPAE*, vol. 1, *The Early Years, 1879-1902*, Princeton, Princeton University Press, 1987.

Stachel John, Cassidy David C., Renn Jürgen, and Schulmann Robert (Eds.), *CPAE*, vol. 2, *The Swiss Years: Writings, 1900-1909*, Princeton, Princeton University Press, 1989.

Klein Martin J., Kox A.J., and Schulmann Robert (Eds.), *CPAE*, vol. 5, *The Swiss years: Correspondence, 1902-1914*, Princeton, Princeton University Press, 1993.

Schulmann Robert, Kox A.J., Janssen Michel, and Illy József (Eds.), *CPAE*, vol. 8, *The Berlin years: correspondence, 1914-1918*, Princeton, Princeton University Press, 1998.

Kormos Buchwald Diana, Illy József, Rosenkranz Ze'ev, and Sauer Tilman (Eds.), *CPAE*, vol. 13, *The Berlin Years: Writings & Correspondence, January 1922-March 1923*, Princeton, Princeton University Press, 2012.

Kormos Buchwald Diana, Rosenkranz Ze'ev, Illy József, Kennefick Daniel J., Kox A. J., Lehmkuhl Dennis, Sauer Tilman, and Nollar James Jennifer (Eds.), *CPAE*, vol. 16, *The Berlin years: writings & correspondence, June 1927-May 1929*, Princeton, Princeton University Press, 2021.

Acknowledgements

I would first like to thank my colleague at the Université Côte d'Azur for more than twenty years, Prof. Jean-Pierre Provost, for our many discussions on Albert Einstein's correspondence with Mileva Marić and Michele Besso, as well as Albert's first scientific writings in 1895. I would also like to thank him for drawing my attention to an article by Max Reinganum, which was essential for understanding the reorientation of Einstein's thesis and more generally for our scientific discussions on Ernst Mach, Ludwig Boltzmann, Paul Drude and Max Planck and the clarifications he provided. Our discussions extended to Albert's environment, as I travelled around Milan and Pavia and gradually discovered new leads. Finally, I would like to thank him for his careful and critical re-reading of this manuscript.

I would like to express my sincere thanks to Prof. Gianpiero Sironi, President of the Lombard Institute, the Academy of Sciences and Letters at the time I was conducting my research, for opening the doors of the Institute to me, and for his availability and encouragement during my work and to Prof. Andrea Silvestri of the Milan Polytechnic, a member of the Institute, for the details he provided during my research, his proofreading and his comments on part of the book.

The work in Pavia owes a great deal to Prof. Lucio Fregonese, President of the Italian Society for the History of Physics and Astronomy, whom I thank for pointing out to me the important documents that Sergio Biscossa's book and a booklet by Fabrizio Bernini represented for my research, introducing me to the staff of the university archives. I would also like to thank Mr Bernini for the information he provided on Ernestina Marangoni's family. I would also like to thank Professor Fabio Bevilacqua, who drew my attention to Albert Einstein's reference to Roberto Schiff.

I would like to express my gratitude to Professor Diana Kormos-Buchwald, Director of the *Einstein Papers Project*, for her invitation to present a preliminary paper as part of the History and Philosophy of Science Seminar at the California Institute of Technology. My thanks also go to Jürgen Renn, Professor of the History and Philosophy of Science at the Max Planck Institute in Berlin, for the references he provided on the genesis of Albert Einstein's idea of light quanta.

The meticulous fieldwork in the archives would not have been possible without the invaluable help of Dr Giuseppina Colombo, head of the mathematics library at the Polytechnic of Milan, Andrea Sottile, of the mathematics library at the University of Milan, whom I thank in particular for his help in identifying the library of the Lombard Institute and Alessandra Baretta and Maria Piera Milani, of the archives at the University of Pavia, for the many archive documents made available. I would also like to thank Dr Yvonne Voegeli, from the archives of the Swiss Federal Institute of Technology in Zurich, for the documents she provided on Vittorio Cantoni and Jakob Rebstein, as well as for our discussions.

I would like to thank Vittorio Cantoni, Professor of Theoretical Physics in the Department of Mathematics at the University of Milan, the eponymous grandson of Vittorio Cantoni, Michele Besso's maternal uncle, who provided me with clarifications and encouragement in the preparatory work for this book. I would also like to thank Valeria Cantoni, great-grand-daughter of Vittorio's younger brother Tullo, who welcomed me on several occasions. I would also like to express my gratitude to Orsa Lumbroso, Director of the Marco Besso Foundation in Rome, for our discussions and, in particular, to Pierre-Luc Besso, Michele's great-grandson, for our talks and for the family details he provided.

I would also like to thank my friends Alessio Moretti and Florence Albrecht for their help with the translations from Italian and German respectively, for which I am responsible (as well as for the English translation of the Einstein-Besso correspondence later than 1928), and my wife, for listening to me every day on this subject for months on end and carefully rereading the manuscript. Many thanks also to my cousin Matelda Lo Fiego for her documentary help and our warm discussions, and to her father Vittorio.

My thanks would not be complete without mentioning Michela Malpangotto, researcher at the CNRS and director of the History of Astronomy team at the Système de Référence Temps-Espace (SYRTE) laboratory at the Paris Observatory, to which I report for my research, for her constant support, including financial support, and Frédérique Vidal, (former) President of Nice Sophia-Antipolis University, for having provided a favourable framework for my research. Finally, a big thank you to Michel Blay, Emeritus Director of Research at the CNRS, without whom this book would not exist. I finally wish to thank again the SYRTE laboratory in Paris Observatory and the National Institute for Teacher Training in Côte d'Azur University, for their financial support in the English translation.

I am profoundly indebted to Alan Rodney for his careful reading of the proofs of this book and his numerous and very useful corrections.

All my thanks to Prof. Dr. Tilman Sauer for his preface to the present book and for his comments.